IMISCOE Research Series

Now accepted for Scopus! Content available on the Scopus site in spring 2021.

This series is the official book series of IMISCOE, the largest network of excellence on migration and diversity in the world. It comprises publications which present empirical and theoretical research on different aspects of international migration. The authors are all specialists, and the publications a rich source of information for researchers and others involved in international migration studies.

The series is published under the editorial supervision of the IMISCOE Editorial Committee which includes leading scholars from all over Europe. The series, which contains more than eighty titles already, is internationally peer reviewed which ensures that the book published in this series continue to present excellent academic standards and scholarly quality. Most of the books are available open access.

For information on how to submit a book proposal, please visit: http://www.imiscoe.org/publications/how-to-submit-a-book-proposal.

More information about this series at http://www.springer.com/series/13502

Alice Massari

Visual Securitization

Humanitarian Representations
and Migration Governance

 Springer

Alice Massari
Sant'Anna School of Advanced Studies
Pisa, Italy

ISSN 2364-4087   ISSN 2364-4095   (electronic)
IMISCOE Research Series
ISBN 978-3-030-71145-0   ISBN 978-3-030-71143-6   (eBook)
https://doi.org/10.1007/978-3-030-71143-6

This Springer imprint is published by the registered company Springer Nature Switzerland AG
The registered company address is: Gewerbestrasse 11, 6330 Cham, Switzerland

To my parents
Giovanna e Paolo

Preface

This book is about the role that images produced by transnational humanitarian nongovernmental organizations (NGOs) play in global governance, with particular emphasis on photographs depicting Syrian displacement between 2015 and 2016. The idea of this research originates in the encounter of my two professional lives. For I combine academic research with professional experience as an active humanitarian practitioner during the response to the Syrian refugee crisis in Jordan between 2013 and 2015 and the prior displacement of Iraqis. This experience within the relief sector has critically affected the drafting of this book. Both as a humanitarian practitioner and as researcher, I have developed ambivalent feelings toward the work of transnational NGOs in the Global South. My personal field experience and studies on the humanitarian system have combined to make me skeptical about humanitarian organizations and their intended goals. I am sympathetic with the large body of scholarship that has pointed out the problems related to the universalist claims of humanitarian organizations (Chandler 2001), the neo-imperialist character of many INGOs interventions (Petras 1999), the complicated relation with politics (De Lauri 2016), the intrinsic culturalism of relief organizations (Cuttitta 2015), and the ambiguous role NGOs play in neoliberal governance policies (Duffield 2001a, 2007).

This notwithstanding these sympathies, I am also strongly convinced of the need for transnational solidarity among people and countries that experience dramatic inequalities, especially when conflict and natural disasters destroy lives. I cannot deny being deeply appreciative of the existence of organizations that work according to an ideal of solidarity and that engage on the front line, providing humanitarian assistance to those affected.

I came to design and carry out this study because of these ambivalent feelings. Humanitarian action is crucial and needed, yet there is large room for improvement. My hope is that my study, as an insider who knows that world, could contribute to its improvement. In this sense, a note of caution is in order. Considering the contemporary socio-political context in which I have written this book, one which is regrettably marked by xenophobic and nationalistic feelings in Europe and elsewhere,

I often find myself in the difficult position of criticizing the "good guys," those who in a particular historic moment were (and still are) trying to affirm positive values of international solidarity in an otherwise rather racist environment. The intention of this book is not to attack humanitarian actors for the sake of criticism. For a world without relief organizations would be even more intolerant, ignorant, closed, and negligent than the one we live in today. It is important, however, given the crucial role now played by international humanitarianism, to reflect on how its discourses and practices affect the attempt to create a better world.

References

Chandler, D. (2001). The road to military humanitarianism: How the human rights NGOs shaped a new humanitarian agenda. *Human Rights Quarterly, 23*(3), 678–700.

Cuttitta, P. (2015). Humanitarianism and migration in the mediterranean borderscape: The Italian-North African border between sea patrols and integration measures. *Borderscaping: Imaginations and practices of border making*, 131–140.

De Lauri, A. (2016). *The politics of humanitarianism. Power, ideology and aid.* IB Tauris.

Duffield, M. (2001a). *Global governance and the new wars: The merging of development and security.* Zed Books Ltd.

Duffield, M. (2007). *Development, security and unending war: Governing the world of peoples.* Polity.

Petras, J. (1999). NGOs: In the service of imperialism. *Journal of Contemporary Asia, 29*(4), 429–440.

Pisa, Italy   Alice Massari

Acknowledgments

It is my great pleasure to thank all the people without whom this book would not have been possible. Starting from my doctoral studies, from which this book began, would like to thank Francesco Strazzari who has shown his interest in my research project and supported its implementation throughout the years. I also owe my deepest gratitude to Anna Triandafyllidou, whose commentary, critique, support, and advice have made this book infinitely better. I hope I have been able to incorporate her broad scope of knowledge and keen analysis in my dissertation.

A special thanks to my friends Mjriam, Monica, Sharif, and Marcello, who have shared their knowledge and sharp criticism with me. They have supported this study in so many ways, hosting me during fieldwork, facilitating networking, forwarding papers, briefs, and artistic works' news relevant to my research, and, most importantly, organizing trips around the world to take a break from the study and enjoy their company. Among them, I would like to especially thank Mjriam, who despite her hectic intellectual and personal life has always found the time to read and comment my work. In my new life in Florence, I am indebted to Serena and Alessandro, who have been the perfect companions for entertainment and academic exchange, and the *Cafecionados* families that have believed in the importance of friendship when social life has been challenged by COVID-19 containment measures.

I am profoundly grateful to my unconditionally supportive family: my sister Mila, my parents Giovanna e Paolo, and my mother-in-law Mariangela who have been there in every moment of need to allow me to work, focus, spend time in the field, attend conferences, and rest, while they were taking care of Mariangela, Pietro, and Antonio. They have travelled last minute by car, train, and plane to make this book possible. They have facilitated relentless travels between Florence, Amman, Milan, San Diego, and so many other places while making the kids feeling always at home. In this endeavor, they have been helped by my extended family, Zio Luigi, Enrico, and Enzo, who have supported in any way they could our crazy life. To all of them goes my gratitude.

I owe a special thank you to my kids Mariangela, Pietro, and Antonio. They have put up with two working and travelling parents, bore my stress, and shared our roaming life around three continents over the past years. Although at times I had been desperate for extra time for my work, I cannot avoid appreciating how their being around, jumping, singing, shouting, and playing has allowed me not to lose my mind in the project and enjoy life with them.

Finally, to Luigi, my life-companion. Words cannot possibly express my gratitude for the unique pleasure of a life spent in his company. To him goes my debt beyond the countless suggestions, comments, and feedback he has given me on this book. We have written, researched, discussed, and travelled together for 14 years by now, and I hope to continue to do so for the rest of our lives.

Note

This book is based on the study of four major transnational NGOs' photographic representation of Syrian displacement between 2015 and 2016: Save the Children, Oxfam, CARE, and Médecins Sans Frontières. I would like to thank MSF, Save the Children, CARE, and the artist Khaled Hourani for their kind permission to use their images. Oxfam did not grant permission.

Contents

List of Figures

Chapter 1
Introduction

"How do we see refugees? The refugee has become a multifaceted symbol, the most prominent political figure of our time" states the brochure of an art retrospective by Khaled Hourani, a Palestinian artist reflecting on the reduction of refugees to abstract symbols of victimhood by humanitarian representations. In the eyes of the artist, the blue figure (Fig. 1.1), so common in relief organizations´ visual depictions, is the migrating human being, without a specific national, religious, ethnic, or gender identity. Yet, the visual landscape of contemporary displacement is anything but abstract. Images of overcrowded boats in the Mediterranean, refugee camps, improvised shelters along migration routes, children and families in need, and people stranded behind fences and walls have come to constitute a powerful reminder of contemporary conditions of displacement for people on the move. Yet, the question remains: how do we see refugees?[1]

This is a book on the role that transnational humanitarian NGOs play in migration governance through visual representation. It offers an innovative account of how relief organizations' visual depiction of Syrian displacement contributes to reproduce and reinforce a securitized account of refugees, one in which refugees are framed in terms of threat. Images of overcrowded boats in the Mediterranean Sea, refugee camps, improvised shelters along the migration routes, children and families in need, and people stranded behind fences and walls have come to constitute a

---

[1] Throughout my work, I will use the term refugees, migrants, displaced people, and people on the move interchangeably. This is, on the one hand, a practical choice since the different terms are often used as synonyms by humanitarian NGOs. On the other, I agree with the studies that have highlighted the (in)appropriateness of the various labels, pointing out how they do not represent the complexities of displacement and they contribute to creating categories of ´deserving´ versus ´undeserving´ asylum seekers. (see among others Zetter 1991, 2007; Long 2013; Crawley and Skleparis 2018). I agree with Monsutti on the need of a "new theoretical sensitivity" critical of "the effects that a national frame of reference may have on research related to refugees and other displaced persons" (Monsutti 2005, 18). For these reasons, and in the attempt to dilute the rigidity of these labels, I will use all terms as synonyms of people on the move. When observing how specific humanitarian organizations use one or other of the terms, I will report it accordingly.

© The Author(s) 2021
A. Massari, *Visual Securitization*, IMISCOE Research Series,
https://doi.org/10.1007/978-3-030-71143-6_1

Fig. 1.1 ©Khaled Hourani, The Blue Figure 2017

powerful reminder of today conditions of global displacement. While visual representation is certainly media-driven today, more than ever, it is humanitarian organizations – and NGOs in particular – who while doing their relief work produce the great bulk of the images presented to the public that contribute to shape our understanding of the refugee issue. In this context, how do transnational humanitarian NGOs represent refugees? What are the implications of this visual depiction within the larger role that relief agencies play in global governance?

The text is based on a multi-method and multi-modal analysis of the role that humanitarian NGOs now play in global governance. The visual investigation focuses on the photographic material representing Syrian displacement published by four major transnational NGOs (CARE, Save the Children, Médecins Sans Frontier, and Oxfam) over the years of the so-called ´European refugee crisis´, between 2015 and 2016. The book analyses a dataset of over 1000 images through a combination of visual methodologies: visual content analysis for a preliminary classification; iconography for its potential in identifying visual trends and patterns in a large body of images, and visual social semiotics for its attention to visual signs and means of representation within a situated cultural and social context. Simultaneously, the exploration of humanitarian visual material has been combined with fieldwork in three of the countries that have hosted, and still host, some of the largest populations of Syrian refugees (Greece, Jordan and Lebanon). Interviews with key informants were combined with analysis of the organization's guidelines and strategic documents.

This book adopts a critical approach to study humanitarianism, humanitarian visuality, and their interrelation with global governance and securitization with the ultimate goal of shedding light on the political, cultural and ethical dynamics at play. It shows that the way through which transnational humanitarian organizations engage in the international political arena is manifold. On the one side, the different NGOs' organizational culture (including mission, vision, approach to humanitarian action, funding strategy and advocacy objectives) shape the distinct roles that each of them has chosen to play in their interaction with state-nation politics and in global governance. On the other side, it empirically demonstrates how relief agencies engage with world affairs through the analysis of their visual production.

In this sense, inspired by the securitization theory elaborated by the Copenhagen school, the book argues that humanitarian NGOs' visual representation of Syrian displacement is contributing to the securitization of the refugee issue. Through visual analysis, it empirically demonstrates how the securitization process takes place in three different ways. First of all, even if marginally, through the reproduction of mainstream media and political accounts that have depicted refugees in terms of threats. Secondly, and more consistently, through a representation of Syrian displaced people that, despite the undeniable innovative aesthetic patterns focusing on dignity and empowerment, nevertheless continues to reinforce a visual narrative around refugees in terms of victimhood and passivity. In so doing, the book does not want to deny the efforts NGOs are currently making to challenge victimizing and essentializing representations of displaced people. On the contrary, they are to be commended. However, there are still many aspects that without further problematization end up impairing their attempts, something this book aspires to untangle. Third, through the dialectic between what is made visible in the picture and what is not, transnational humanitarian NGOs reinforce a representation of Syrian refugees that reduces the portrayed individuals to abstract humanitarian subjects, eventually eliding any political claim and agency that Syrian people on the move might have.

The book argues that there is also another important dynamic at play. While on the visual level NGOs continue to securitize and depoliticize Syrian refugees (despite dramatic recent changes in the way humanitarian information is communicated), the analysis of NGOs' policy and strategic documents, and the interviews with practitioners have also revealed a countertrend. Indeed, relief agencies put a great deal of effort into trying to empower the beneficiaries of the assistance and also to go beyond the traditional humanitarian communication that – as humanitarian actors they are themselves perfectly aware of – unfairly depicts refugees in terms of victimhood and helplessness. In this sense, despite the consistency of contemporary visual representations in portraying refugees with no (or very limited) agency, it also seems that the continuous intellectual engagement and efforts put into changing this depiction is creating what Yurchak defines as the "minute internal displacements and mutations into the discursive regime in which they are articulated" (Yurchak 2013, 28).

These micro mutations are visible to a certain extent in the wider range of aesthetic topics that some NGOs present to the public. For example, when the represented subjects encompass the actual threats looming over people in their countries, in the hosting countries and along the migration routes (such as perpetrators of violence, border apparatus, law enforcement, fences, and detention centres), and the complexity of displacement condition not only in terms of humanitarian need, but also in terms of everyday life, community solidarity, and mundane details not necessarily linked to the experience of loss. It is in expanding the visual patterns used to portray refugees into a photographic portfolio that humanitarian organizations have the opportunity to offer a discourse counterpoising the mainstream, one able to account for the complexities and multiple narrative facets of displacement.

1.1   Contributions

This book intends to make four important academic contributions. First, it sets out to expand the body of literature on migration and securitization by providing a thorough study of visual representational practices of transnational humanitarian NGOs. The analysis will not only touch upon themes such as the changes in aesthetic patterns over time, but will also highlight similarities and differences among the four major relief organizations' visual registries and narratives. Scholars have shown the different ways in which humanitarianism and securitization interplay (inter alia see Aradau 2004a; Andersson 2014; Musarò 2017; Chouliaraki and Georgiou 2017). A few have pointed out the role that relief organizations play as actors of securitization (Aradau 2004a; Watson 2011). However, none of the existing studies have demonstrated empirically how these dynamics work. This book intends to fill this gap by showing how humanitarian NGOs contribute to the securitization of the refugee issue through their visual representations.

Secondly, in the field of International Relations (IR), nobody has systematically carried out an analysis using visual social semiotics. More generally, as Lene Hansen – one of the scholars who has produced the most ground-breaking work on visual securitization – has noticed, very few studies in IR have engaged in the investigation of large bodies of images. Most existing studies have employed content analysis that focuses on the identification of the portrayed subjects. However, none of these works have explored how these forms of content analysis differ or could be combined with other visual methodologies (Hansen forthcoming). Moreover, as Andersen et al. have observed the "most recent 'visual security' work in IR has been on the iconic image and has assumed a 'powerful intertextuality around the image' which puts the image 'in danger of vanishing" (Andersen et al. cited in Hansen forthcoming, 8). This book seeks to address these challenges in three ways. It is based on a large body of images (over a thousand). It combines the content analysis with a more qualitative visual methodology, and it addresses the need to attribute more attention to the image *per se.*

This study draws on an additional innovative methodological approach. It is not only based on a multi-methods analysis that combines different visual methodologies, but it is also a multi-modal investigation. Thorough visual analysis has been combined with fieldwork in three of the countries that have hosted, and still host, some of the largest populations of Syrian refugees. Since most academic work on images and security has been based on content analysis, by combining it with visual social semiotics, and especially by incorporating fieldwork, this study intends to offer a different understanding of the role and impact of images within the wider context of global governance.

Last, but not least, this book links visuality with governance. This is particularly interesting when one considers that humanitarian communication and governance interplay at different levels. Relief organizations have at certain historical times worked in competition with nation-states and international policies, such as in the case of the aid blockade during the Biafra crisis, and more recently during search and rescue (SAR) operations in the Mediterranean which contrast with European migration policies. At the same time, NGOs' the funding and advocacy strategies of many humanitarian are tightly connected with states' political agenda. In today's highly competitive media environment aid agencies represent the refugees at the same time that they represent themselves, their mission, and goals. In this sense, transnational humanitarian organizations, and their communication practices, merit study because they do political work in a political environment. This book unpacks the dynamics through which relief agencies contribute to the humanitarian systems of governmentality through their visual communication and shows empirically how these two dimensions of visuality and governance do relate to each other and interact.

1.2 Why ´How Do We See Refugees´ Matters

Questions around the representation of refugees and the implications of visual narratives in the understanding and management of displacement are particularly relevant in the context of contemporary political debate on the ´refugee crisis´. It should be noted that I use the term critically because, as many scholars have pointed out, the intensity of arrivals in Europe has neither been a new phenomenon in the history of migration, nor has it constituted the ´invasion´ that has been presented in public accounts (see among others Fargues 2015; De Genova et al. 2016; Krzyżanowski et al. 2018). The large majority of refugees are still hosted by Syria's neighbors. It is undeniable that the term ´refugee crisis´ has come to define a specific historical phenomenon that entails important political implications in the way refugees are represented, perceived, and ultimately managed and how societies in countries of destination are deciding to face contemporary questions of international mobility. In today's securitized context (Krzyżanowski et al. 2018; Mountz 2015; Huysmans 2016) it is therefore particularly interesting to explore the role played by some of the most important actors in the management of the refugee crisis – the humanitarian organizations – which are not commonly associated with the securitization effort routinely performed by other entities (such as Frontex, law enforcement agencies or national border apparatuses).

All this is especially thought-provoking if we consider the role that humanitarian NGOs have been performing in terms of humanitarian governance (Fassin 2011; Barnett 2013) not only in the Global South (e.g. Syria's neighboring countries), but also at the very centre of the Mediterranean vis-à-vis refugees, European migration policies, and domestic politics. The highly politicized and mediatized debate around refugees' mobility (Krzyżanowski et al. 2018) has been further intensified by discussions around maritime SAR and accusations made by populist governments that NGOs facilitate irregular migration. NGOs have produced research reports reflecting on the impact of their maritime operations and concluded that not only are the accusations unsubstantiated, but that the "involvement of humanitarian vessels was associated with a significant improvement in maritime safety compared to other periods" (MSF 2017b). Even though legal accusations may, at least for the moment, have diminished (Repubblica 2018b), statal institutions (especially the Italian Ministry of Interior and the Libyan coast guard) were – at the time of writing – preventing SAR operations from happening (Repubblica 2018a; Rome 2017; Cuttitta 2018). The topic remains therefore very actual and relevant as public political debate rages and political scientists have shown how NGOs with SAR operations have contributed to the re-politicization of international waters into a political environment from which they can advocate for humanitarian commitment, political change in migration management, and solidarity (Stierl 2018; Cuttitta 2017), while at the same time they themselves "become part of a hybrid border management system" (Cuttitta 2017, 20).

From a visual perspective the odyssey of people attempting to find better lives in third countries has attracted unprecedented public attention with the publication of sadly famous pictures such as that of Alan Kurdi, the three-year-old Syrian whose lifeless body was washed up on the Turkish shore after his inflatable boat capsized. Such images, together with those of rubber dinghies full of people arriving on Greek shores, those of child victims of chemical gas attacks and of Syrian cities bombed and besieged have animated national and international public debate. This study is important because it pays special attention to the visual representation of displacement and it unpacks the implications that humanitarian visuality has on the refugee issue. For focus on humanitarian visuality "can help address a gap in visual studies, which in its analyses of the myriad of significant actors shaping the ways in which images are created, selected, displayed, and interpreted – whether we think of national governments, private corporations, media organizations, political movements and campaigns, or lay audiences – has tended to overlook the crucial role of NGOs in informing the visual cultures of national and global civil societies" (Kurasawa 2015, 44).

1.3   Research Design

1.3.1   The Syrian Crisis

This book is based on the analysis of the visual representation of Syrian refugees offered by four major transnational humanitarian NGOs (CARE, Save the Children, Oxfam and Médecins Sans Frontières) between 2015 and 2016. It looks particularly at the images produced and disseminated online through the agencies' websites and Facebook pages.

Since the beginning of the conflict in 2012, Syrian people have sought refuge in neighboring countries. At the time of the research, over 5.5 million people were registered as Syrian refugees with the United Nations High Commissioner for Refugees (UNHCR) (UNHCR 2017c). Turkey hosted the biggest proportion (nearly three and a half million people), Lebanon registered a million refugees, Jordan

655,000, Iraq 230,000 (in addition to 3.2 million internally displaced people), and Egypt around 115,000 (UNHCR 2017c). The humanitarian effort inside Syria has been severely hindered by the very limited access to affected populations, dramatic level of violence, and serious security concerns for humanitarian staff. Most relief has been delivered in neighboring host countries where UN agencies, international organizations and most transnational NGOs have established or substantially enlarged their operations. Starting from March 2012, the coordination of the humanitarian response has been systematized at the regional level under the leadership of UNHCR, as specified under the UN-wide cluster system.[2] The number of agencies that participated in the Humanitarian Regional Response Plan rose from 34 in 2012 (UNHCR 2012) to 144 in 2017 (UNHCR 2017a). Organizations active in the response to the Syrian crisis mostly include UN agencies, transnational and local NGOs, and a few foundations. Over the years, relief agencies have been working in all sectors of humanitarian interventions (such as health, education, shelter and protection). Their regional aggregated budget has risen from $84,159,188 in 2012 (UNHCR 2012) to $4,400,570,955 in 2017 (UNHCR 2017a).

In 2015, the movement of Syrian people took another direction. For a variety of reasons, including the worsening of the situation in several neighboring countries (Achilli 2015; UN 2015), and opening of the eastern Mediterranean and Balkan route (Fargues 2015), Syrians also started to move toward Europe, where a little less than a million people applied for asylum (UNHCR 2017b). The intensity of arrivals of people of different nationalities – but mostly from Syria – on Europe's southern coasts, and especially on Greek shores, between 2015 and 2016 led many humanitarian, institutional and academic commentators to talk about a refugee, migrant or Mediterranean ´crisis´ (ECHO 2015; UNHCR 2015; Fargues 2015; Pallister-Wilkins 2016). The monthly number of sea arrivals in Europe increased from around 5000 people in January 2015 to a peak of over 220,000 in October 2015, after which it gradually decreased until the end of 2016 (UNHCR 2017d). At the end of 2016 the number of arrivals in Europe drastically decreased, *de facto* concluding the crisis. What happened was the entry into force in March of that year of an agreement between the European Union and Turkey, the so called EU-Turkey Agreement (European Council 2016). The deal aimed at stopping irregular migration flows from Turkey to Europe by preventing Syrians (and others on the move) from travelling from Turkey to Greece or from Greek islands to the Greek mainland and then onwards into Europe and also by forcibly returning those judged to have moved irregularly.

Together with this new strict migration regime European discourse on migration became increasingly militarized and securitized (Perkowsky 2016; Garelli and Tazzioli 2017; Musarò 2017). This study focuses specifically on this period, from

---

[2] The Cluster approach is a coordination mechanism (set by the United Nations General Assembly Resolution 46/182 in 1991 and defined in the 2005 Humanitarian Reform Agenda) which seeks to improve the coordination of the international humanitarian response. Clusters are coordination groups divided into thematic areas, including protection, water and sanitation, nutrition and health) whose members include major aid agencies working in the respective sectors.

January 2015 to December 2016, a time whose seemingly unexpected association of humanitarianism and securitization discourses in the response to the ´crisis´, provides a unique opportunity to empirically explore their interconnected and mutually constitutive nature (Aradau 2004a).

1.4   The Non-governmental Organizations

Within the variegated landscape of humanitarian organizations responding to the Syrian crisis, I have had to circumscribe the object of analysis. Acknowledging that humanitarianism has undeniably various expressions originating from different historical and cultural backgrounds (Kennedy 2005; Fiori 2013), I also agree with Pacitto and Fiddian-Qasmiyeh (2013). They have noted how dominant humanitarian theory has been based on a tradition of Western humanitarianism with its specific set of values and practices. Present-day humanitarianism – defined by some authors as a 'new humanitarianism'– with its special relationship with the political sphere and its assumingly universal language, is even more embedded in Western culture (Chandler 2001). Without denying the relevance of ´other´ important forms of relief action, my book will focus on the tradition of Western professionalized and transnational humanitarianism.

Among the wide range of different organizations operating in the humanitarian sector, I decided to focus on non-governmental organizations, as opposed to UN agencies and state actors, for two main reasons. First, because although it is a fact that humanitarian assistance is today provided by a multiplicity of actors, NGOs have nowadays become one of the most prominent actors of the humanitarian sphere, performing "the vast majority of the operating work of the international humanitarian system" (Maxwell and Walker 2014, 117). Secondly, and most importantly, a large body of scholarship has explored the relationship between International Organizations (IOs) and securitization (see for example the work of Abrahamsen 2005; Andersson 2012; Chandler 2007; Gabiam 2016; Geiger and Pécoud 2014; Gerstl 2010; Russo and Giusti 2017; Longo 2013; Mason 2014; Musarò 2013, 2017). Analysis of the securitizing role of NGOs has been much less investigated (with the notable excptions of Duffield 2007, 2014; Aradau 2004a; Watson 2011).

The world of transnational humanitarian NGOs is still populated by a numerous and variegated set of agencies. The term NGO encompasses organizations with different geographical membership and activities. There are national organizations that only operate in a specific country and transnational ones that intervene across borders and are composed of a variety of members in different States. This study concentrates on the latter because of their wider geographical presence, outreach and potential to shape the transnational humanitarian discourse. Among all organizations, I decided to focus on the major ones. Cottle and Nolan (2007) describe the Red Cross, Save the Children, Oxfam, World Vision, CARE and Médecins Sans Frontières (MSF) as "the world's major aid agencies" (Cottle and Nolan 2007, 862) and show how these organizations have contributed to shape the discourse of today's humanitarianism. Similarly, Barnett and Weiss (2008) list MSF, CARE, S, World Vision International, Catholic Relief Services (CRS), and Oxfam as the best-known humanitarian agencies. Among these organizations, I have selected the four NGOs identified by the authors quoted above (i.e. CARE, MSF, Oxfam and Save the Children). Moreover, since the time-space framework of the research is the humanitarian response to the Syrian refugee crisis both in the neighboring countries and in Europe between 2015 and 2016, this study will focus on the organizations that have or had operations in all those areas. This will inevitably exclude World Vision and CRS as they have not participated in the refugees/migrants emergency response in the Mediterranean. The International Committee of the Red Cross (ICRC) has been excluded because, although often referred to as an NGO, it has in fact a hybrid status.[3]

1.5 Multi-modal Analysis

The case study is based on a multi-modal analysis that combines the use of multiple methods, namely: visual methodologies (visual social semiotics and iconography) and multi-sited fieldwork. As Bleiker has argued, the "political significance of images is best understood through an interdisciplinary framework that relies on multiple methods" (Bleiker 2015, 872). Since images circulate and acquire meaning at various overlapping levels, simultaneously in the international arena and in the viewer's mind, Bleiker has observed how their meaning is always dependent on context and interpretation. For this important reason the methodology of this book is based on the combination of visual methods with a multi-sited fieldwork in three countries along with analysis of NGOs' strategic documents, policies, and guidelines.

1.5.1 Images Collection and Classification

The visual data used for this study consist of 1184 pictures. Images have been collected from the websites and Facebook pages of the four selected NGOs according to the following criteria: (1) images should be photographic images; (2) they should represent Syrian displaced people and/or the causes of their displacement and (3) they should have been published between 2015 and 2016. Data have been stored on Nvivo software for classification and analysis purposes. For websites, I privileged the pictures published on the NGOs international profile as opposed to the websites of the different country offices. Given the impossibility of navigating websites

---

[3] ICRC is neither exactly an NGO nor an Intergovernmental Organization because it is not mandated by any Government. However, it is treated like other intergovernmental organizations and enjoys privileges reserved for this latter category.

historically, I have collected images using *the way back machine* free software that periodically takes snapshots of websites' different pages and stores them online. Through this tool, it has been possible to access NGOs' website pages from January 2015. Every relevant page was saved in PDF and HTML format every 4 months until December 2016. All pictures have been individually uploaded on Nvivo.

For Facebook pages, the international account has been taken into consideration, or the account specifically dedicated to the Syrian crisis. When the latter was not available, as in the case of Save the Children and Oxfam, the choice has fallen on the profile of the organization headquarters. Data were automatically extrapolated by Nvivo software through its application for social media analysis that creates datasets listing all posts, pictures, video, and metadata of the selected period. I eliminated the duplicates, keeping the image the first time they were used within a media. In the case of pictures used both on Facebook and an agency's website, unless specifically required by the image (e.g., publication with a particular caption, of for belonging to a Facebook campaign), I privileged data found on websites where the compositional meaning could also be explored. Since the focus of the study is the transnational humanitarian NGOs' representation of Syrian displacement, I have not considered the images that would only depict NGOs staff, though I kept the pictures representing NGOs workers and managers with Syrian refugees. The data collection of pictures from the websites and Facebook pages of the four selected NGOs has generated a corpus of 1184 pictures composed as indicated in the table below (Table 1.1).

1.5.2   Multi-sited Fieldwork

Field research for this study was carried out intermittently between March and November 2017. Through multi-sited ethnography and semi-structured interviews with key stakeholders, I gathered a ´visual reconnaissance´ of the locations of the so-called Syrian refugee crisis. This opened a window into how visual products are conceived and developed, the NGOs' intentions and key messages behind the representation of the Syrian crisis, and practitioners' understanding of their respective organizations' visual communication policies and strategies. The presence and movements of both Syrian refugees and major humanitarian NGOs in Syria's neighboring countries and in the Mediterranean makes a multi-sited research the most appropriate method of enquiry. I decided to focus on Lebanon and Jordan in the Middle East, and Greece on the European shore where Oxfam, SCF, CARE and

Table 1.1 Breakdown of data collected, divided by producing NGO and media

	CARE	OXFAM	MSF	SCF	Total
Website Syria 2015–2016	113	110	357	155	735
Facebook Syria 2015–2016	295	32	31	91	449
Total	408	142	388	246	1184

MSF have ongoing humanitarian operations. For each of the four organizations on which this study focuses, I have gathered and analysed organizational guidelines, strategic documents, and policy briefs. In all locations I visited, project implementation sites, formal and informal refugee camps, I interviewed NGO staff, humanitarian advocacy and communication specialists, local authorities, photographers working for the NGOs, volunteers active in the emergency response, and academic researchers.

1.6   Methodological Considerations

A couple of methodological considerations are in order before proceeding. The first concerns my professional experience as a humanitarian practitioner. Although I am surely not the first researcher to have worked in the humanitarian sector before undertaking an academic career, I am inspired by a post-modern sensibility. I believe that to go beyond the object-subject dichotomy and the myth of the ´view from nowhere´ it is crucial to reflect on the researcher's positionality. The best way to acknowledge the situated nature of my research (Finlay and Gough 2008) is to address the question of self-reflexivity. In this sense, I cannot deny that my experience in the field, and particularly my work in responding to the Iraqi displacement following the US-led invasion in 2003 and the Syrian refugee emergency caused by the conflict in Syria, have influenced my analysis. Although ultimately interested in contributing with my work to the solidarity effort undertaken by many organizations, I found myself increasingly reflecting on the nature of information flows within the sector in which I worked.

My research has also been a way to explore how some humanitarian practices were also causing more harm than good. Probably more often than I wanted, my eyes were at first looking for what was wrong about an image, taking for granted what was indeed good. And that is also why I had frequently to go back in my analysis and make more explicit the different meanings of the pictures so as to clearly acknowledge what I always expected to be included in the image: the ´good´ message. Eventually, my position toward the object of analysis and its situated nature, helped me to account for the polysemic character of images. Systematically reflecting and making sense of it throughout the research has been essential.

A second consideration regards the use of visual material. In showing with pictures what I was arguing I may have risked falling into the trap of inadvertently implying that what was presented visually was therefore ´true´ or somehow ´real´, a premise that I discard epistemologically in recognizing that all visual representation – photography included – constitutes an interpretation of reality and not its objective visual transposition. Drawing on Bal's concept of expository tradition (1996), Szörényi (2006) notices how by presenting pictures of refugees, academic publications can reproduce the problematic dynamic that they were instead trying to challenge. That made me think that although throughout the study I have sought to provide alternative interpretation of the images presented, and to highlight the absence of the voice of the people represented in the pictures, my position toward the represented participants remains that of an outsider, one that does not give any voice to them, their interpretation, and their self-representation.

My last reflection concerns the methodological strength of much visual analysis. While acknowledging the existence both of extremely important contributions specifically focusing on the methodological aspect of visual analysis (Rose 2001; Van Leeuwen and Jewitt 2001; Hansen 2011; Heck and Schlag 2013) and also the fundamental advancement at this level, there may still be room for improvement. In many excellent IR and political science studies on visuality I read I could not help being struck that frequently the paper omitted to indicate and explain the methodology used for the visual analysis and the elaboration of the argument. This is not a minor flaw. For, especially in the relatively new field of visuality in IR, it ultimately undermines the power of the insight and the perspicacity of the argument. Against, this background, I hope that with the extensive effort put into the definition of a visual methodology from both a theoretical and practical perspective, I have been able to improve the strength of my insight.

1.7   The Content of this Book

The book is divided into six chapters. Following the introduction, Chap. 2 presents the theoretical framework that constitutes the backbone of the whole study. After introducing the notion of humanitarianism and presenting the academic discussion around the interrelation between humanitarianism and politics, it outlines the role that civil society plays in global governance with a particular attention to the specific role of transnational humanitarian NGOs. In so doing, the section links present-day humanitarianism, sometimes referred to as ´new-humanitarianism´, with governance. In the second part, the chapter presents the analytical framework of the securitization theory elaborated by the International Security Studies (ISS) literature. It explains how the approach of the Copenhagen School is the one which is here adopted to explore securitization, one of the realms in which relief organizations actively participate in the international political arena and global governance. The section also outlines the different studies that have so far focused on the interrelation of humanitarianism and securitization and clarifies how this study is innovative, for the first time focusing on visuality, instead of practices. Finally, the last part, puts the work of transnational humanitarian NGOs into the contemporary context of communication and the highly competitive environment in which organizations contend for funds, public mobilization and media attention. This section touches upon the features of humanitarian communication, and how it has changed over time, with particular attention to the humanitarian representation of refugees.

Chapter 3 completes the theoretical framework by providing the intellectual premises of a visual approach to the study of IR. It frames visual analysis within the larger strand of critical discourse analysis to then explain the origin of the attention given to images and the added value of visual social semiotics – combined with iconography – as method of enquiry. Since this book is based on photographic images, this section also outlines what are the features and the political-epistemological claims of this specific genre and addresses the crucial aspect of the polysemic value of images.

Chapters 3, 4, 5 and 6 constitute the analytical part of the book. Chapter 3 is dedicated to the analysis of the specific role that each of the four selected NGOs plays in global governance. The exploration is based on the investigation of each of the relief organizations' background, humanitarian approach, mission, vision, funding policy, and advocacy strategy. The investigation of the four different organizational models allows us to highlight similarities and differences in the ways each of the four NGOs selected for this study perceives, describes and presents its own role within global governance.

Chapter 4 is the first of the following three chapters based on the visual analysis. It focuses on the different NGO representational practices that contribute to the depiction of Syrian people on the move as a threat. Through the visual analysis of five aesthetic patterns – vectors without goal, emergency, boats, conceptual structures, and the visual construction of the other – this section intends to shed light on the ways through which aid agencies, to a certain degree, reproduce and reinforce a culturally and politically dominant public and media securitized account of refugees.

Chapter 5 is the exploration of images that, in the attempt to challenge mainstream discourses of securitization around people on the move, privilege a representation of refugees that portrays them as referent object of a threat. In line with contemporary humanitarianism's focus on human security, protection and the rights-based approach, these kinds of depictions introduce innovative elements and perspective. This section is dedicated to the exploration of six aesthetic patterns – pity, victimization, infantilization, suffering, innocence, and the savior-hero - that consistently represent Syrian displaced people securitized as referent objects of different kinds of existential risks. The aim is to show how these relatively new kinds of depiction continue to portray refugees in terms of victimhood, passivity and very limited agency.

The last of the analytical chapters, Chap. 6, focuses on invisibility, the absences that emerged from the visual analysis and the comparison between humanitarian (discursive) narrative and the elements visually present in the NGOs' pictures. It approaches the representation of Syrian refugees with a particular attention to the dialectic between what is visible in the images and what is not, and the exploration of the various reasons behind absence. The chapter aims to show how various securitization processes are connected with visibility but also, and not less importantly, with invisibility. It argues that paying attention to this dialectic is important to glimpse glimmers of minor displacements and the way NGOs represent refugees. Finally, the conclusion sums up the various findings and shows how they empirically support the argument that transnational humanitarian organizations play a key role in global governance, specifically in the way in which their accounts of displacement interrelate with mainstream securitization discourses. However, within this visual depiction it is possible to distinguish signals of micro-displacements.

References

Abrahamsen, R. (2005). Blair's Africa: The politics of securitization and fear. *Alternatives, 30*(1), 55–80.

Achilli, L. (2015). *Syrian refugees in Jordan: A reality check*. Policy Brief, MPC, EUI, 2015. http://cadmus.eui.eu/bitstream/handle/1814/34904/MPC_2015-02_PB.pdf

Andersson, R. (2012). A game of risk: Boat migration and the business of bordering Europe (Respond to this article at Http://Www.Therai.Org.Uk/at/Debate). *Anthropology Today, 28*(6), 7–11.

Andersson, R. (2014). *Illegality, Inc.: Clandestine migration and the business of Bordering Europe*. University of California Press.

Aradau, C. (2004a). Security and the democratic scene: Desecuritization and emancipation. *Journal of International Relations and Development, 7*(4), 388–413.

Bal, M. (1996). *Double exposures: The subject of cultural analysis*. Psychology Press.

Barnett. (2013). Humanitarian governance. *Annual Review of Political Science, 16*, 379–398.

Barnett, M., & Weiss, T. G. (2008). *Humanitarianism in question: Politics, power, ethics*. Cornell University Press.

Bleiker, R. (2015). Pluralist methods for visual global politics. *Millennium, 43*(3), 872–890.

Chandler, D. (2001). The road to military humanitarianism: How the human rights NGOs shaped a new humanitarian agenda. *Human Rights Quarterly, 23*(3), 678–700.

Chandler, D. (2007). The security–development Nexus and the rise of "anti-foreign policy". *Journal of International Relations and Development, 10*(4), 362–386.

Chouliaraki, L., & Georgiou, M. (2017). Hospitability: The communicative architecture of humanitarian securitization at Europe's Borders. *Journal of Communication, 67*(2), 159–180.

Cottle, S., & Nolan, D. (2007). Global humanitarianism and the changing aid-media field: "Everyone was dying for footage". *Journalism Studies, 8*(6), 862–878.

Crawley, H., & Skleparis, D. (2018). Refugees, migrants, neither, both: Categorical fetishism and the politics of bounding in Europe's "migration crisis". *Journal of Ethnic and Migration Studies, 44*(1), 48–64.

Cuttitta, P. (2017). Repoliticization through search and rescue? Humanitarian NGOs and migration management in the Central Mediterranean. *Geopolitics*, 1–29.

Cuttitta, P. (2018). *Pushing migrants Back to Libya, persecuting rescue NGOs: The end of the humanitarian turn (Part I)*. Oxford Law Faculty, April. https://www.law.ox.ac.uk/research-subject-groups/centre-criminology/centreborder-criminologies/blog/2018/04/pushing-migrants

De Genova, N., Tazzioli, M., Álvarez-Velasco, S., De Genova, N., Heller, C., Peano, I., Riedner, L., Bermant, L. S., Spathopoulou, A., & Suffee, Z. (2016). Europe/crisis: New keywords of "the crisis" in and of "Europe". *New Keywords Collective, Near Futures Online*, 1–45.

Duffield, M. (2007). *Development, security and unending war: Governing the world of peoples*. Polity.

Duffield, M. (2014). *Global governance and the new wars: The merging of development and security*. Zed Books Ltd..

ECHO. (2015, October 28). *Refugee crisis*. https://web.archive.org/web/20151115014133/http://ec.europa.eu/echo/refugee-crisis_en

European Council. (2016). *EU-Turkey statement, 18 March 2016 - Consilium*. http://www.consilium.europa.eu/en/press/press-releases/2016/03/18/eu-turkey-statement/

Fargues, P. (2015). *2015: The year we mistook refugees for invaders*. Policy Brief, MPC, EUI, 2015. http://cadmus.eui.eu/bitstream/handle/1814/38307/Policy_Brief_2015_12.pdf

Fassin, D. (2011). *Humanitarian reason: A moral history of the present*. University of California Press.

Finlay, L., & Gough, B. (2008). *Reflexivity: A practical guide for researchers in health and social sciences*. Wiley.

Fiori, J. E. M. (2013). *Humanitarian affairs think tank*.

Gabiam, N. (2016). Humanitarianism, development, and security in the 21st century: Lessons from the Syrian refugee crisis. *International Journal of Middle East Studies, 48*(2), 382–386.

Garelli, G., & Tazzioli, M. (2017). The biopolitical warfare on migrants: EU naval force and NATO operations of migration government in the Mediterranean. *Critical Military Studies*, 1–20.

Geiger, M., & Pécoud, A. (2014). International organisations and the politics of migration. *Journal of Ethnic and Migration Studies, 40*(6), 865–887.

Gerstl, A. (2010, September 9–11). The depoliticization and'ASEANization'of human security in Southeast Asia: ASEAN's counter-terrorism and climate change policies. In *Working paper, prepared for standing group on international relations, 7th Pan European international relations conference*. Stockholm.

Hansen, L. (2011). Theorizing the image for security studies: Visual securitization and the Muhammad cartoon crisis. *European Journal of International Relations, 17*(1), 51–74.

Hansen, L. (forthcoming). Visual international relations: Ontologies, methodologies and migration across the Mediterranean. *Internatioanl Studies Quarterly*.

Heck, A., & Schlag, G. (2013). Securitizing images: The female body and the war in Afghanistan. *European Journal of International Relations, 19*(4), 891–913.

Huysmans, J. (2016). 13 migration and the politics of security. *Minorities in European Cities: The Dynamics of Social Integration and Social Exclusion at the Neighbourhood Level, 179*.

Kennedy, D. (2005). *The dark sides of virtue: Reassessing international humanitarianism*. Princeton University Press.

Krzyżanowski, M., Triandafyllidou, A., & Wodak, R.. (2018). *The mediatization and the politicization of the "refugee crisis" in Europe*.

Kurasawa, F. (2015). How does humanitarian visuality work? A conceptual toolkit for a sociology of iconic suffering. *Sociologica, 9*(1), 0–0.

Long, K. (2013). When refugees stopped being migrants: Movement, labour and humanitarian protection. *Migration Studies, 1*(1), 4–26.

Longo, F. (2013). The relevance of security sector reform in humanitarian intervention: The case of the European Union in the Mediterranean. *Democracy and Security, 9*(1–2), 177–192.

Mason, M. (2014). Climate insecurity in (post) conflict areas: The biopolitics of United Nations vulnerability assessments. *Geopolitics, 19*(4), 806–828.

Maxwell, D. G., & Walker, P. (2014). *Shaping the humanitarian world*. Routledge.

Monsutti, A. (2005). *War and migration: Social networks and economic strategies of the Hazaras of Afghanistan*. Routledge.

Mountz, A. (2015). In/Visibility and the securitization of migration: Shaping publics through border enforcement on islands. *Public Culture, 11*(2), 184–200.

MSF. (2017b). *DEFENDING HUMANITY AT SEA*.

Musarò, P. (2013). " Africans" vs. "Europeans": Humanitarian narratives and the moral geography of the world. *Sociologia Della Comunicazione*.

Musarò, P. (2017). Mare nostrum: The visual politics of a military-humanitarian operation in the Mediterranean Sea. *Media, Culture & Society, 39*(1), 11–28.

Pacitto, J., & Fiddian-Qasmiyeh, E. (2013). *Writing the 'Other'into humanitarian discourse: Framing theory and practice in South–South humanitarian responses to forced displacement* (RSC Working Paper Series 93).

Pallister-Wilkins, P. (2016). Interrogating the Mediterranean "migration crisis"'. *Mediterranean Politics, 21*(2), 311–315.

Perkowsky, N. (2016). Deaths, interventions, humanitarianism and human rights in the Mediterranean "migration crisis". *Mediterranean Politics, 21*(2), 331–335.

Repubblica. (2018a). Aquarius, no di Italia e Malta. Si offrono Barcellona, Napoli e Palermo. *Repubblica.it*. http://www.repubblica.it/cronaca/2018/08/13/news/aquarius_e_il_soccorso_rifiutato-204002144/

Repubblica. (2018b). Palermo, Archiviate Le Indagini Sulle Ong: "Nessun Legame Coi Trafficanti Di Migranti" - Repubblica.It. http://palermo.repubblica.it/cronaca/2018/06/19/news/palermo_archiviate_le_indagini_sulle_ong_nessun_legame_coi_trafficanti_di_migranti_-199435698/

Rome, Jon Henley Angela Giuffrida in. (2017). Three NGOs halt Mediterranean migrant rescues after Libyan hostility. *The Guardian*, 14 August 2017, sec. World news. https://www.theguardian.com/world/2017/aug/14/three-ngos-halt-mediterranean-migrant-rescues-after-libyan-hostility

Rose, G. (2001). *Visual methodologies: An introduction to researching with visual materials*. Sage.

Russo, A., & Giusti, S. (2017). *Monuments under attack: From protection to securitisation*. http://cadmus.eui.eu//handle/1814/47164

Stierl, M. (2018). A fleet of mediterranean border humanitarians. *Antipode, 50*(3), 704–724.

Szörényi, A. (2006). The images speak for themselves? Reading refugee coffee-table books. *Visual Studies, 21*(01), 24–41.

UN. (2015). *VULNERABILITY ASSESSMENT OF SYRIAN REFUGEES IN LEBANON 2015*. https://reliefweb.int/sites/reliefweb.int/files/resources/2015VASyR.pdf

UNHCR. (2012). *Syria Regional Response Plan*. file:///C:/Users/Alice/Downloads/RRP1-March2012.pdf

UNHCR. (2017a). *3RP Regional Refugee & Resilience Plan 2018–2019*. file:///C:/Users/Alice/Downloads/3RPRegionalStrategicOverview2018-19.pdf

UNHCR. (2017b). *Syria regional refugee response*. UNHCR Syria Regional Refugee Response. http://data.unhcr.org/syrianrefugees/asylum.php

UNHCR. (2017c). *Syria regional refugee response - Regional overview*. http://data.unhcr.org/syrianrefugees/regional.php

UNHCR. (2017d, June 30). *Situation Mediterranean - Sea Arrivals*. https://data2.unhcr.org/en/situations/mediterranean#

UNHCR, United Nations High Commissioner for. (2015, July 1). *Mediterranean crisis 2015 at six months: Refugee and migrant numbers highest on record*. UNHCR. http://www.unhcr.org/news/press/2015/7/5592b9b36/mediterranean-crisis-2015-six-months-refugee-migrant-numbers-highest-record.html

Van Leeuwen, T., & Jewitt, C. (2001). *The handbook of visual analysis*. Sage.

Watson, S. (2011). The 'Human'as referent object? Humanitarianism as securitization. *Security Dialogue, 42*(1), 3–20.

Yurchak, A. (2013). *Everything was forever, until it was no more: The last soviet generation*. Princeton University Press.

Zetter, R. (1991). Labelling refugees: Forming and transforming a bureaucratic identity. *Journal of Refugee Studies, 4*(1), 39–62.

Zetter, R. (2007). More labels, fewer refugees: Remaking the refugee label in an era of globalization. *Journal of Refugee Studies, 20*(2), 172–192.

Part I
The Theory and Methodology of Visual Securitization

Part I
The Theory and Methodology of Social
Stratification

Chapter 2
Humanitarianism, Securitization and Humanitarian Communication

2.1 Introduction

To unpack the role that transnational humanitarian NGOs play in contemporary systems of governance and highlight how they contribute, through their visual production, to the securitization of the refugee issue, it is important to introduce the notions of humanitarianism, global governance, and securitization. Also, since NGOs do not operate in a vacuum but within a highly competitive media environment, it is equally important to reflect on the ways in which humanitarian ideals are translated into their communication strategies and how these fit within the larger communication landscape. This chapter presents the literature and the theoretical framework on which this book is based.

It starts by briefly presenting the concept of humanitarianism, its origins and characteristics, by also addressing the 'elephant in the room', the ambiguous relationship that humanitarianism entertains with politics. Since the beginning of humanitarian assistance its relationship with politics has been subject to debate. If it is true that the notion of humanitarianism is based on concepts and principles that mostly have to do with the realm of ethics and also very often defined in opposition to the political dimension, its relationship with politics has always been very complex. A brief overview of the debate around this interplay is essential as the need to understand this complex relation is an assumption underlying the whole study.

Going beyond the intense debate on the interconnections between humanitarianism and politics, this book focuses on the role that NGOs perform in global governance. The second section of the chapter lays out the literature that has highlighted the position of civil society in influencing the supra-national political arena and the key role that transnational organizations play in it. For an exploration of the specific ways through which contemporary transnational humanitarian NGOs engage in global politics and global governance, it is also necessary to present the features of present-day humanitarianism (sometimes referred to as 'new humanitarianism'). Academic work has investigated the various mechanisms through which relief

© The Author(s) 2021
A. Massari, *Visual Securitization*, IMISCOE Research Series,
https://doi.org/10.1007/978-3-030-71143-6_2

agencies seek to affect policy changes in the various national and international contexts in which they intervene. This study builds on this strand of work to explore how the different organizations' origins, missions, policies, and approaches to humanitarianism indicate different dynamics of participation in global governance.

Among the different sectors in which NGOs can play a role in today's global governance (such as on issues around economy, environment, and health) this study is particularly interested in the security dimension. This is mainly because the refugee issue has been consistently linked with security discourses by various political and media accounts. In apparent contrast, the work of humanitarian NGOs has been commonly associated with an ethical and moral dimension. The third section hence outlines the securitization theory developed by the Copenhagen School. Its theoretical approach is all the more valuable as it has explained how security can be intended as something objective, but that it also has a very important discursive dimension. The concept of securitization is particularly relevant insofar as it enables us to grasp different dimensions of the securitization process. We thus see not only the framing of refugees in terms of threat, but also their framing as terms of referent objects of a threat with the notion of human security. Inspired by the International Security Study literature, I build on the body of scholarship that has focused on the interconnections between securitization and humanitarianism to empirically show how NGOs have contributed to the securitization of the refugee issue.

To do so it is also important to look at transnational humanitarian organizations' communication policies and visual material. Humanitarian communication is the object of the last section of this chapter. In order to fully grasp its impact on the constitution of humanitarian discourse(s), it is crucial to understand that NGOs do not operate in a vacuum. Rather, relief organizations contend, in a highly competitive environment, for funds, public support, and media attention. Moreover, communication strategies, representational styles and approaches need also to be contextualized culturally and historically. Clearly, they assume different meanings in different cultures, and over the years have undergone substantial modification. The chapter concludes with the presentation of how refugees have been visually represented in humanitarian communication throughout history.

2.2 Humanitarianism and Its Complicated Relationship with Politics

Humanitarianism, as a form of compassion and charity, has been present throughout history. Since its origin, the practice of humanitarianism has had different forms depending on the various cultural, religious or secular, philosophical and philanthropic inspirations of the concept. Although the idea of saving lives and alleviating suffering has been present in different cultural and religious backgrounds, "modern humanitarianism's origins are located in Western history and Christian thought" (Barnett and Weiss 2008, 7). The concept – in the sense it is used today – finds its origin in the early nineteenth century and became, acknowledged and, in some ways, institutionalized, by the creation of the International Committee of the Red Cross (ICRC) in 1863 (Barnett 2011).

Barnett and Weiss' (2008) understanding of humanitarianism helps us seems adept in grasp the socially constructed and historically contingent meaning of the term. The notion is based on the way practitioners have conceptualized the humanitarian sphere since its early usages. "Many within the humanitarian sector tend to conceive the ideal humanitarian act as motivated by an altruistic desire to provide life-saving relief; to honour the principles of humanity, neutrality, impartiality, and independence; and to do more good than harm" (Barnett and Weiss 2008, 10–11). More formally, the United Nations Office for the Coordination of Humanitarian Affairs (OCHA) defines humanitarian assistance as "aid that seeks, to save lives and alleviate suffering of a crisis affected population" (OCHA 2003, 13).

Despite the fact that the concept of humanitarian action is a rather complex notion, defined in different ways by different humanitarian actors (Collinson and Elhawary 2012), there is a general consensus considering neutrality, impartiality and independence as the fundamental principles of humanitarian actions (Leader 2000; OCHA 2003).[1] Today the majority of humanitarian actors have embraced the same humanitarian principles in their official documents (Hilhorst and Jansen 2010). Consequently, according to OCHA, the humanitarian space is a "conducive humanitarian operating environment" in which agencies can work in compliance with the principles of neutrality and impartiality and their intervention can be independent from military and political action (OCHA 2003, 32).

Within the humanitarian space there are a multiplicity of actors operating in different capacities, including Governments' aid structures – like for example the United States Agency for International Development (USAID) or the former British Department for International Development (DfID) (renamed in 2020 as the Foreign, Commonwealth & Development Office) – International Organizations (IOs) – such as the United Nations (UN) agencies and the EU Agency for Humanitarian Aid and Civil Protection (ECHO) – organizations with hybrid status – i.e. the International Committee of the Red Cross (ICRC) and NGOs. Among this assorted group of agencies, this study focuses on the NGOs which, that as Stoddard has noticed, "have evolved into a crucial pillar of the international humanitarian architecture." (Stoddard 2003, 1). What substantially differentiates NGOs from all other humanitarian actors is summarized in the first part of their acronym: non-governmental. In this sense, NGOs constitute themselves as actors completely independent from State politics and interests and often, in contraposition to these, as actors willing and able to take action when States seem negligent, absent or, worse, the perpetrator of the wrongdoing which need to be addressed.

---

[1] The principles of neutrality, impartiality and independence, along with another four fundamental principles, are defined in the Statutes of the International Red Cross and Red Crescent Movement, adopted by the 25th International Conference of the Red Cross, Geneva (ICRC), 1986, available on the ICRC website. For an overview of the basic principles of humanitarianism and their practical implications, see (Slim 2015).

The qualifier 'non-governmental', combined with the adherence to humanitarian principles, are the two elements that situate humanitarian NGOs in a humanitarian space, in theory unrelated with politics and power but actually, on the contrary, constituting fundamental features of governments' interests. Despite the aspiration of humanitarianism to neutrality and independence (Hilhorst and Jansen 2010), a wide range of scholarship has highlighted how humanitarian agencies, including NGOs, have a much more complex relationship with politics (among others Weiss 1999; Duffield 2001; Slim 2003; Rieff 2003; Donini et al. 2004; Barnett and Snyder 2008). A better understanding of the multifaceted and heterogeneous relationships that NGOs have with politics and power could help us to make sense of the complex role that humanitarian organizations play in current global governance and security.

Humanitarianism has been defined as situated in the realm of ethics (Barnett 2013); a substitute for politics (Higgins 1993); the essence of politics (Cutts 1998; Weiss 1999) or the politics of life (Fassin 2007). Debate on the implications of the interconnections between humanitarianism and politics is far from over.[2] Within the wide range of perspectives, the two scholarly positions at the extremities of the spectrum are worth mentioning. In his seminal *A Bed for the Night*, David Rieff (Rieff 2003) argues that traditional humanitarian action has been ruined by politicization. Rieff's claim for a politically independent humanitarianism is based on the opinion that any connection with politics and long-term goals such as the promotion of human rights and good governance have basically reduced and hindered the humanitarian space at the expense of the victims. On the contrary, argues Slim, not only "humanitarianism is always politicized somehow. It is a political project in a political world. Its mission is a political one – to restrain and ameliorate the use of organised violence in human relations and to engage with power in order to do so" (Slim 2003, 1). For the author, the politicization of humanitarianism does not pose a problem in itself. Rather, it is a matter of choosing the 'good' politics of humanitarianism.

The centrality of the problematic relationship of humanitarianism with politics, intended mainly as government politics at the national and inter-governmental level, is reflected in different scholarly attempts to categorize NGOs. Even though extremely heterogenous in the definition of variables, all typologies hitherto proposed in the literature have considered the level of interrelations with politics (in its various forms) as a central part of the equation. In 1999, Weiss classified NGOs along two axes. On one side, he considered the strengths of the relationship of the organization' humanitarian action with politics, and, on the other, the NGOs' attitude toward "traditional operating principles" such as neutrality, impartiality and non-discriminatory provision of aid. According to Weiss (1999), ICRC represents the perfect example of 'classicist' organization and its humanitarian action as completely separated from politics, with no engagement with political authorities. At the other end of the spectrum Weiss situates MSF, an agency which for the author,

---

[2] (For different position on the complex relationship between humanitarianism and politics see Cutts 1998; Dany 2015; De Lauri 2016; Donini et al. 2006; Duffield et al. 2001; Fiori 2013; Hilhorst and Jansen 2010; Macrae and Leader 2001; Nascimento 2015; Terry 2013; Fox 2001);

embodies the "solidarist" organization, an NGO that considers abandoning neutrality to pick sides in disputes to be part of its organizational mandate.

In reviewing their trends and challenges Stoddard (2003) categorizes NGOs by their Dunantist or Wilsonian traditions. Dunantist is derived from Henri Dunant, founder of the Red Cross and winner of the first Nobel Peace Prize, and refers to the importance given to the principle of independence. The Wilsonian approach, named after the American President Woodrow Wilson, refers to its hope to "to project US values and influence as a force for good in the world" (Stoddard 2003, 2). The author, considering NGOs' attitudes more or less dependent on and cooperative with governments, classifies CARE, Save the Children US and IRC as Wilsonian and Oxfam, Save the Children UK and MSF as Dunantist.

Dijkzeul and Moke (2005) have created a typology based on two axes. On the horizontal, the authors consider the principles according to which aid is delivered: "impartial" if the organization provides non-discriminatory and need-based assistance and based on "solidarity" if the agency contemplates siding with one of the parties to the conflict and/or their political cause. On the vertical axis, NGOs are classified according to the relationship that they entertain with States and particularly through their level of dependence on institutional donors in the country where their headquarters are situated, in the case of major UK NGOs the former DfID and for US NGOs with USAID. According to the typology developed by Dijkzeul and Moke, MSF and ICRC are organizations with the highest degree of impartiality and independence, followed by Oxfam and Save the Children UK who appear slightly below but within the same matrix's quadrant. CARE, at the bottom of the left quadrant, is classified as an organization with a moderate level of impartiality but strongly dependent on State funding.

In another classification effort, Hansen (2007) mapped humanitarian organizations working in Iraq according to criteria that included the "degree of cooperation with/distance from political and military actors, degree of visible acceptance/refusal of security from combatants or security contractors, and acceptance/refusal of funding from parties to the conflict" (Hansen 2007, 68). Save the Children UK, Oxfam GB and MSF have been classified in the Dunantist/Contractors quadrant, while CARE appeared on the right border of the same quadrant, leaning toward the Wilsonian/Pragmatist end of the axis.

Barnett and Snyder (2008) have highlighted how, with the exception of ICRC and MSF,- almost all of the other NGOs "have accepted the challenge of attempting to transform societies and have become comfortable with their political intentions" (Barnett and Snyder 2008, 158). In a subsequent and longer publication Barnett (2009) proposes a more detailed map of the organizational responses based on the different relationships of the humanitarian agencies with politics, identifying two main approaches. On one hand, there are "emergency" organizations such as MSF that are primarily concerned with relief assistance and are generally more independent. On the other hand, NGOs such as CARE International can be defined "alchemic" because of their transformative aspiration to social change while addressing the root causes of suffering which are generally more dependent on States. Even though alchemic organizations could be accused of being more implicated with politics, they usually disagree and instead present their political interventions in technical terms. Chapter 4 will explore this dynamic in detail.

Far from offering a homogeneous indication about the different NGOs position vis-à-vis politics and levels of (in)dependence from governments, what all these different classifications suggest is that idea of the humanitarian space as an environment completely unrelated with politics and power needs to be problematized. So does the idea of the purist compliance with humanitarian principles since all these studies have highlighted that, in one way or another, NGOs need to accommodate and negotiate those principles in order to achieve their strategic objectives and fulfill their specific mandates. Although it would be extremely interesting to directly engage with the intense scholarly debate on the interrelations between humanitarianism and politics, this is outside the scope of this study. What I believe it is important to retain here is an understanding of humanitarian action as strictly and, in various ways, interrelated with power and state politics at different levels. In fact, not only do NGOs do work in a highly politicized context (especially during conflicts) but also they carry out work that has, and very often is intended to have, important political implications. As Duffield has effectively summarized, the politics of humanitarian action: "in the sense that humanitarian action is 'political', would appear to hinge on two factors. First, it circumscribes all the decisions, actions, compromises, and so on, that humanitarian actors make during the course of their work. Second, these decisions and actions are political in that they are seen as making a difference and capable of altering outcomes" (Duffield 2001, 96). In the following sections I will attempt to define more clearly how exactly relief agencies interact with politics and power specifically through the lenses of their role within global governance and security.

2.3 NGOs and Global Governance

2.3.1 NGOs and Global Governance: A General Overview

Scholars have long acknowledged that transnational NGOs have today become one of the key non-state actors of global governance (see, for example, Weiss and Gordenker 1996; Duffield 2001, 2007; Chandler 2004). As Weiss (2000) has noticed, the scholarly use of the term global governance is mostly based on the observation that although non-state actors are not something new in the international system, their proliferation, combined with their increasing relevance and power, has become a distinguishing character of present-day international affairs. Rosenau's understanding of global governance, intended as "systems of rule at all levels of human activity—from the family to the international organization—in which the pursuit of goals through the exercise of control has transnational repercussions" (Rosenau 1995, 13) is particularly useful. For it allows us to consider social practices as able to affect the economic, social and environmental spheres even in the absence of formal institutions authorized to take action (Oran Young cited in Weiss 2000).

In discussing global governance – and the role of transnational humanitarian NGOs within it – I draw on the work of Barnett and Duvall on power in global governance (Barnett and Duvall 2004). The two authors point out that the two terms are indeed strictly linked as governance encompasses the norms, structures and organizations that rule and control social life. In their understanding, power refers to the creation of effects that, through social relations, affect the ability of actors to shape their state and prospects. Among the four different expressions of power that they identify in *Power in Global Governance* – compulsory, institutional, structural and productive – the notion of productive power is particularly relevant for this book. "Productive power is the socially diffuse production of subjectivity in systems of meaning and signification" (Barnett and Duvall 2004, 3). In Barnett and Duvall's taxonomy of forms of power productive power is *diffuse* – it works through connections at a certain 'distance' or that are mediated –and works through *social relations of constitution* – social relations that determine who the actors are and what capacities they have, their 'power to'. Productive power is therefore about "discourse, the social processes and the systems of knowledge through which meaning is produced, fixed, lived, experienced, and transformed" (Barnett and Duvall 2004, 20). Discourses are here intended in Foucauldian terms as sites of social relations of power that produce social identities and capacities. Barnett and Duval's conceptualization is particularly relevant here as it allows analysis of the discursive production – including contested ones – of subjects and the determination of meanings in global governance. In particular, it allows us to investigate how the 'other' is defined and how the fixing of meaning is connected with determined practices and policies (Barnett and Duvall 2004).

The added value of this approach is that it enables a study that not only looks at how a specific actor regulates the world, but also at how the actors constitute it. In order to explore the role played by transnational humanitarian NGOs in global governance, it can be useful to situate the discussion within the broader conceptual framework of the role of transnational civil society in global governance. The syncretic definition of civil society provided by Edwards (2009) seems to be particularly apt to acknowledge the concept as a "contested territory" (Edwards 2009, vii), and conveys the complexity of the term. In his definition, the concept of civil society merges three theoretical models that see civil society respectively as the associational/non-profit sector, a normative model of 'good' society, and as an arena for public debate and action. Even though the origins of the concept of civil society are to be found in the Greek philosophical tradition, the modern notion of civil society separated from the State was articulated by Hegel in the nineteenth century to be then defined further as the public space of political, ideological and cultural debate in the twentieth century (Kaldor 2003). Present-day civil society does still share a "common core of meanings" (Kaldor 2003, 585) with previous definitions. Its contemporary feature is that of having transcended national borders to become transnational (Florini 2012; Kaldor 1999) or even global (Anheier et al. 2001; Kaldor 2003). The actors of transnational civil society are heterogeneous and include social movements and advocacy networks (Davies 2014), coalitions and various kinds of associations (Florini 2012), grass-root groups (Anheier et al. 2001), and, incontestably, non-governmental organizations (Anheier et al. 2001; Chandler 2004; Davies 2014; Florini 2012; Kaldor 1999, 2003).

2.3.2 Humanitarian NGOs Role in Global Governance

While the key role that NGOs plays in global governance has been widely acknowledged, it is important to consider how relief organizations specifically perform this role, and what are the distinctive aspects of present-day humanitarianism that define them as actors in global governance. The next three sections intend to highlight the characteristics of contemporary humanitarianism by shedding light on the different ways each influences and attempts to transform the supra-national political arena.

2.3.2.1 The Features of Contemporary Humanitarianism or 'New Humanitarianism'

Present day humanitarianism has often been referred to as "new humanitarianism" (see among others Tirman 2003; Fox 2001; Chandler 2001; Duffield 2001; Macrae 2002). Duffield et al. (2001) places its origin in the 1990s, at a time when it was being realised that traditional humanitarian action was often perversely having unintended negative consequences. This realisation led to the emergence from a coalition of NGOs of the Do no Harm (DNH) principle which has subsequently become enshrined in the guiding principles of most humanitarian agencies.

There is no universally accepted definition of the concept of 'new humanitarianism'[3] but its characteristics can be traced in the humanitarian sector's self-acknowledgement of the complex relation of their actions with politics (and often the explicit decision to directly engage with this reality as part of the organization's mandate). A second feature has to do with the increasingly common merging of development intervention into humanitarian action. The last distinctive character of the new humanitarianism is the widespread (although not always accepted, as I shall show below), shift from a needs-based approach to humanitarian assistance to a rights-based approach.

With regards to the first feature – the increasingly explicit interrelation of humanitarianism and politics – in the 1990s there was a shift of perspective from the traditional focus of humanitarian action on 'only' saving lives and alleviating suffering. As Macrae and Leader explained "during the 1990s a consensus emerged within the international humanitarian system that there was a need to enhance the 'coherence'

---

[3] (For a compelling review of different authors' attempts to provide a definition of new humanitarianism, see Tirman 2003)

between humanitarian and political responses to complex political emergencies. Closer integration between aid and political responses was seen to be necessary in order to address the root causes of conflict-induced crises, and to ensure that aid did not exacerbate political tensions" (Macrae and Leader 2001, 290). Duffield has pointed out how "the new humanitarianism, therefore, implies a drive for coherence where humanitarian action becomes part of a comprehensive political strategy" (Duffield et al. 2001, 271). Academic perspectives regarding the coherence approach are far from homogenous. Scholars are divided between those who have welcomed the approach, all things considered, as a positive development (Abiew 2003; Charny 2004; Harmer 2008; Slim 2003; Weiss 1999 among others) and those who have underlined its challenges and remain extremely critic (see for example Donini et al. 2008; Macrae and Leader 2001; Metcalfe et al. 2012; Pandolfi 2003). Leaving aside this interesting debate, the importance of this relatively new distinctive feature is that not only is humanitarian action conceptualized and implemented as part of international political endeavors, but, most importantly, relief agencies become key actors when it comes to response to (Macrae and Leader 2001) or management of (Dillon and Reid 2000) contemporary complex emergencies.

The second distinctive character of new humanitarianism is the integration of long-term objectives into the relief intervention (Duffield 2001), in other words, the merging of humanitarian action with development. To understand what this means, it is probably important to briefly go back to the definitions of the two terms. While humanitarian action is mostly intended to address the immediate need of an emergency situation, such as providing urgent health care, food and water, and emergency shelter, development work has a long-term time frame, usually applied in non-emergency settings and aiming at addressing conditions of under-development. Development includes a much larger set of reforming interventions that can include infrastructure, vocational training, female empowerment, rule-of-law promotion, judicial reform or agriculture and education support. The decision to integrate both humanitarian and development objectives into a sort of holistic approach to a crisis situation is therefore a move with important implications for the kind of impact humanitarian actors have within the societies in which they operate. Although, as I have discussed earlier, humanitarianism does not exist in a realm extraneous to politics and power, the link between development, power and politics seems even more explicit. Seminal academic contributions have clearly highlighted how development policies have worked as systems of control. Their extensive and pervasive effects in the countries of their operations are similar to the mechanisms of colonialism (Escobar 2011). As a neo-colonial Western project, it has often hidden the root causes of underdevelopment amidst its obsession with economic growth (Rist 2014). As Duffield et al. (2001) has argued, the new humanitarianism has become development-oriented with an aspiration to transform society and to avoid conflict which is seen as among the effects of underdevelopment. Development discourse, traditionally seen as depoliticized and technocratic, reveals its politicization exactly through its intentions of radical social change. The new or politicized humanitarianism "complements the radicalisation of development which now sees the role of aid as altering the balance of power between social groups in the interests of peace and

stability. From saving lives, the shift in humanitarian policy has been towards analysing consequences and supporting social processes" (Duffield 2001, 80).
Consequently, decisions regarding the extent of organizational mandates, whether
to stick to a purely humanitarian concern to relieve suffering or one or which combines relief with development, have practical political implications for the way an
organization perceives its role in global governance. The different NGOs' choices
in this regard and their implications are discussed in Chap. 3.

The last major feature of the new humanitarianism lies in its approach to humanitarian assistance. While early humanitarianism has been traditionally based on a
needs-based approach, intended to save lives and alleviate people's suffering
according to their practical' needs, over the last decades there has been a shift
toward a new paradigm: the right-based approach. Contemporary humanitarian aid
not only targets people's needs, but it also focuses on protecting people's rights.
New humanitarianism is more concerned with the outcomes of the assistance, rather
than the assistance *per se* (Duffield 2001). The shift toward this new paradigm has
not been accepted by practitioners and scholars without controversy. The critique
has included post-colonial (Chandler 2001; Žižek 2005) and feminist (Brems 1997;
Coomaraswamy 1994) critiques the human rights regime *tout-court*, adding to an
intense debate on the ethics, effectiveness and righteousness of the rights-based, as
opposed to the needs-based, approach (Brown 2014; DuBois 2007; Duffield 2014;
Fox 2001; Rieff 2003; Slim 2003). Chandler has pointed out "the transformation of
humanitarianism from the margins to the center of the international policy agenda
has been achieved through the redefinition of humanitarian policy and practice and
its integration within the fast-growing agenda of human rights. The new international discourse of human rights activism no longer separates the spheres of strategic state and international aid from humanitarianism but attempts to integrate the
two under the rubric of "ethical" or "moral" foreign policy" (Chandler 2001, 678).

2.3.2.2 Advocacy

Following the end of the Cold War NGOs started to use public advocacy as part of
their strategies to accomplish their humanitarian mandate. Although the role of
advocacy in the history of humanitarianism is not completely new, it has progressively gained momentum. Today, the importance of influencing, campaigning or
advocating for something – according to the different organizations' definitions – in
relief organizations' work has increased to such a point that many NGOs, in their
programmatic documents, mention advocacy on the same level as their 'hands-on'
humanitarian and development interventions. The increasing attention NGOs pay to
advocacy has been also complemented by moves towards greater cohesion and the
coordinated networks such as the International Council of Voluntary Agencies
(ICVA) or the Voluntary Organizations in Cooperation in Emergencies (VOICE).
Through them aid agencies can more effectively influence national and international
policies, (Duffield 2001). This influence, as Hudson has specified, seeks policy
change "to favour the poor and marginalized Southern communities whose interests

NGOs aim to promote" (Hudson 2001, 33). Advocacy is, therefore, one of the practical tools through which NGOs participate in global governance by explicitly trying to influence policy-making at different local, national and global levels. Transnational NGOs advocacy networks lobby on a wide range of topics including international debt reduction, child labor, landmines, and the arms trade (Hudson 2001). Particularly, humanitarian advocacy has been interested in drawing attention to forgotten humanitarian crises and complex political emergencies and progressively introduced long-term development objectives (such as fighting poverty or empowering women) so as to address the root causes of underdevelopment.

Present day humanitarian advocacy has been read in connection with the rights-based approach of new humanitarianism. There are opposed views, either seen as a positive development (Booth 1991b) or as something intrinsically problematic (Chandler 2002; Pugh 1998; Pupavac 2006). For its advocates, such 'moral advocacy' is a sign of the development of a global social movement based on moral and ethical human rights discourse. As Booth has argued "Universal human rights are solidly embedded in multiple networks of cross-cutting universal ethical communities. The fundamental weakness of the critics of universality is that they take too territorial a view of the idea of human community, human political solidarity and human social affinity. Their perspective is conservative, overdisciplined by constructed notions of states and cultures" (Booth 1991b, 61).

For its critics, the increasing focus that humanitarian organizations have put on advocacy has contributed to the erosion of a more principled humanitarianism and consequently subordinated people's needs to strategic human rights objectives (Pugh 1998). Moreover, as Pupavac (2006) has argued, for most NGOs the calls to intervention to protect people since the 1990s have been accompanied not only by aspirations to promote social change but also by openly challenging the sovereignty of the developing state in question when state actors were considered as responsible or complicit in violence. With this logic, humanitarian advocacy has tended "to reinforce international inequalities rather than overturning them, by casting conditions in the developing world as moral rather than political and material issues, with dubious results for those in whose name the advocacy is conducted."(Pupavac 2006, 268). There are opposed scholarly views on the appropriateness and righteousness of contemporary humanitarian advocacy. In this debate what I believe is extremely relevant is acknowledgement of the importance that relief organizations' advocacy work perform into global governance. How this happens in practice will be explored in Chaps. 4, 5, 6 and 7.

2.3.2.3 Humanitarian Governance

The key role played by humanitarian actors in the global arena over the last decades has led one of the most famous scholars of humanitarian studies to coin the term 'humanitarian governance' (Barnett 2013). Drawing on the work of Michael Foucault and Didier Fassin, Barnett has defined the modern international humanitarian order in terms of 'humanitarian governance': a global system of governance of humanity aimed at the protection of lives, the alleviation of suffering and the general improvement of individuals' well-being. The notion encompasses a wide range of activities pertaining to the humanitarian sphere, including relief, security, nutrition, development and medical assistance. Inspired by a critical theoretic tradition, and including power in his study of governance, Barnett has suggested ways to reorient the exploration of humanitarian governance in different important ways (Barnett 2013; Barnett and Duvall 2004). First, he suggested addressing attention to the *effects* of an humanitarian enterprise, rather than the study of its effectiveness. He further rejects the assumption that humanitarianism is inherently good, acknowledging that it can serve to promote both emancipation and domination. He has advocated for incorporation of the study of power into the study of humanitarian governance. Additionally, he has sought to shift focus from the normative characteristics of the system to questions about its construction. Finally, he highlights the importance of studying the "underlying discourses of humanitarianism themselves" (Barnett 2013, 382) as constitutive humanitarian actors. All of these points have inspired this book, for they allow us to unpack the role that transnational humanitarian NGOs, and their underlying discourses, play in global governance.

As Barnett has noted, academic study of the different aspects of governance has often been treated in terms of governmentality (Dean 2009 and Neumann & Sending 2006 cited in Barnett 2013). This was defined by Foucault as the "ensemble formed by the institutions, procedures, analyses, and reflections, the calculations and tactics that allow the exercise of this very specific albeit complex form of power"(Foucault 1991, 102). That is also why it is all the more important to take into consideration power in general and Barnett and Duval's notion of productive power in particular. In this perspective, non-state actors are no longer considered in opposition to state power, but rather a "most central feature of how power operates in late modern society" (Sending and Neumann 2006, 652). The literature on humanitarian governance has identified several ways through which humanitarian actors influence governance. These include rationalization processes that present political and power dynamics as bureaucratic and knowledge-based systems (Ferguson 1990); bureaucratization (Barnett and Duvall 2004); psychosocial support (Pupavac 2001b); gender equality (Olivius 2014); biopolitics (Reid 2010; Rozakou 2012; Vaughan-Williams 2015); the international refugee regime (Barnett 2002) as agents of stability in areas of instability (Duffield et al. 2001), and as key actors in the management of complex emergencies (Dillon and Reid 2000). Within this extremely interesting strand of studies, I am particularly interested in those which have focused on the way relief agencies participate in global governance in connection with international security, and securitization in particular. In order to speak about security, a key concept of international relations, and a crucial part of today's debate not only on global governance, but also on humanitarianism, development, and migration it is important to take the time to introduce it theoretically. The next section is dedicated to the definition of the concept and the relatively more recent innovative conceptualization of securitization.

2.4 Securitization and Humanitarianism

2.4.1 Security, Securitization and the Copenhagen School

From a theoretical perspective, this book builds upon the International Security Studies (ISS) literature that has broadened the concept of security and elaborated the notion of securitization. In their ground-breaking *Security: A New Framework for Analysis*, Buzan et al. (1998) criticized the realist ontological and epistemological approach to security that considered the State the (only) unit of analysis within the international security system. The call for widening the study of security beyond the traditional focus on States, already introduced by Krause and Williams (1996), has allowed for analysis of other subjects as referent objects of security and the consideration of security threats other than military force. The second crucial contribution of *New Framework* has been the elaboration of a concept able to broaden the notion of security without an invalidating "conceptual stretching" (Sartori 1970): the concept of securitization. This fundamental notion refers to the framing of an issue in terms of security. It is a discursive construction in which "a securitizing actor uses a rhetoric of existential threat and thereby takes an issue out of what under those conditions is 'normal politics'" (Buzan et al. 1998, 24–25).

Wæver (1995) has conceptualized securitization as a rhetorical structure by virtue of which the actor asserts the urgency of the situation and the consequential need for exceptional measures. According to its theorists, in order to have a case of securitization, there must be the acceptance of a determined discourse. Without it, we can only talk of a 'securitizing move'. However, the acceptance does not refer in any way to the implementation of the security measures implied in the specific issue's framing. Rather, it is about the level of resonance that a securitization discourse is able to attain in legitimizing emergency measures. Therefore, as Buzan and Wæver have argued, the role of the researcher is to investigate the "process of constructing a shared understanding of what is to be considered and collectively responded to as a threat" (Buzan et al. 1998, 26). In its first conceptualization, securitization has been defined as a speech act, "the utterance itself that is the act. By saying the words, something is done"(Buzan et al. 1998, 26). Although in the *New Framework* the authors in the book specifically referred to language theory, the possibility and the necessity of attention to other utterance dimensions, with particular emphasis on the visual realm, but not only it, has been widely discussed (Hansen 2011; Heck and Schlag 2013; McDonald 2008; Williams 2003). I will go back to the visual approach in the next chapter. Here, I would like to focus on the fundamental value that this theoretical framework offers for this book.

The theory of securitization – elaborated by Buzan and Wæver and subsequently to become known as the Copenhagen School – is one of the most innovative and productive areas of research in international security studies (Huysmans 1998; Williams 2003), In many ways it is relevant to this book. First of all, by broadening the security agenda it has underlined the importance of studying the role that non-state actors play in international relations theory in general and in the field of security studies in particular. Secondly, by opening up the possibility that securitization actors are to be found beyond traditional military and other state agents, it offers a theoretical framework to investigate securitization discourses and practices of actors not formally attributable to security apparatuses such as humanitarian organizations. Indeed, in the elaboration of the concept of securitization the Copenhagen School was concerned with understanding what could be the thresholds of the notion in terms of scale and significance in order not to lose its powerful theoretical utility. Instead of trying to identify different levels of importance of the various security referent objects, it suggested the need to focus on the impact of the securitization process. Instead of asking how important is a sector to have more 'securitization' potential, it proposed considering "how big an impact does the securitizing move have on wider patterns of relations? A securitizing move can easily upset orders of mutual accommodation among units" (Buzan et al. 1998, 26).

Williams (2003) has pointed out that while broadening the range of securitizing actors, the Copenhagen School has delineated its borders by identifying the specific structure of the securitization process. It can, in principle, come from 'any' actor that can in turn frame any issue and any referent object in terms of security. At the same time, the securitizing claim has different levels of efficacy that depend on the "external, contextual and social" (Buzan et al. 1998, 32) and the position of the actor vis-à-vis the issue and the audience. In this sense, the analysis of humanitarian NGOs securitizing potential is particularly interesting exactly because of the role they play in humanitarian discourses, a domain seemingly antithetical to security utterances. As Barnett and Weiss (2008) have shown, aid agencies derive power primarily from their alleged expert and moral authority.[4] Borrowing form Dahl's definition of power as "effects" (Barnett and Weiss 2008, 40) that determine the ability of other actors to impact circumstances, the authors highlighted how humanitarian organizations have power in the sense that they affect regulatory and constitutive effects. On the one hand, humanitarian actors seek to govern other's actions through both normative and symbolic practices. On the other, constitutive effect sheds light on the reality as social construction and determines the boundaries of normality, desirability, and best solutions. In this sense, the exploration of humanitarian NGOs discourses and their securitization potential is even more interesting if we take into account their supposed expert and moral authority.

---

[4]As Barnett and Weiss (2008) have pointed out "expert authority exists when an actor's voice is given credibility because of his or her specialized training, knowledge, or experience. Moral authority exists when an actor is perceived to be speaking and acting on behalf of the community's values and interests and defending the lives of the weak and vulnerable" (Barnett and Weiss 2008, 39).

2.4.2 Securitization, Societal Security and Human Security

In broadening the agenda of critical security studies, the Copenhagen School has identified five security sectors, each one with specific threat and referent objects: the military, environmental, economic, societal and the political (Buzan et al. 1998). Within this framework, the concept of societal security elaborated by Wæver is particularly relevant for this book. It refers to the security of the society in terms of identity and as opposed to State security. For him "society is about *identity*, the *self-conception* of communities, and those individuals who identify themselves as *members* of a particular community. *"Society"* should basically be conceived of as both *Gemeinshaft* and *Geselshaft*, but thereby, to some degree, necessarily more than the sum of the parts (that is, not reducible to individuals)" (Wæver 1995, 66).

In its original conceptualization, societal security referred mainly to European societies that securitizing processes framed as referent object of threats. However, the notion is applicable to any society. This was and is crucial because for the first time it included the individual in the picture of the international security system as traditionally conceived. Successively, in a book on the evolution of International Security Studies, Buzan and Hansen (2009) discussed the widening and deepening of the security agenda. The authors presented the different approaches calling for the need to expanding the concept of referent object of security beyond the Western State and suggested including human security as referent object and development as security sector. They argued that 'UNDP's [United Nations Development Programme] conceptualization of conceptualization of Human Security is probably the most encompassing expansion of the concept since Galtung launched structural violence and, like Marxist Peace Research, it sought to bring development and North–South issues into ISS.' (Buzan and Hansen 2009, 203).

Given the importance that the concept of human security[5] has in discourses about the protection of people's lives, and its conceptual origin – historically situated in the dramatic genocides of Rwanda and former Yugoslavia in the 1990s – approaching it through the lenses of securitization theory brings into discussion the question of ethics, particularly the ethics of securitization. While other perspectives on securitization have seen human security as potentially emancipatory (see in this regards the position of what has come to be known as the Welsh School and particularly the work of its most prominent advocates Booth 1991a; Wyn Jones 1999), the Copenhagen School has not attributed to the securitization process any intrinsic positive acceptation.

According to the Welsh School, the concept of human security, by placing the individual at the heart of the international security system, implies an emancipatory promise. In a prophetic speech delivered at the British International Studies Association in 1991, right at the end of the Cold War, Booth argued that in the new world order security meant the absence of threat. In his words, "emancipation is the

---

[5] For a definition of the concept of human security, see Massari, A. (2020a) *Human Security* In Humanitarianism: Keywords (91–93). Brill.

freeing of people (as individuals and groups) from those physical and human constraints which stop them carrying out what they would freely choose to do. War and the threat of war is one of those constraints, together with poverty, poor education, political oppression and so on. Security and emancipation are two sides of the same coin." (Booth 1991a, 319). For Booth, human security is of crucial importance because of its pivotal role in a State-based international security system. For threats to human beings and their rights create instability at the national level that could easily have global spill-over effects. In this conceptualization, security would encompass both security concerns based on domestic interests and those based on the need to fight oppression and underdevelopment. Although given attention mainly for its functional role in a traditional security framework, human security is already linked with the concept of emancipation at the point where for Booth, at the theoretical level, emancipation means security.

In further development of this critical security studies approach, the Welsh School has conceptualized the two concepts so intertwined that in the words of Wyn Jones "security in the sense of the absence of the threat of (involuntary) pain, fear, hunger, and poverty is an essential element in the struggle for emancipation" (Wyn Jones 1999, 126). Since then, scholars have infused human security with claims of tremendous emancipatory power in different sectors. These include the potential to challenge economic, social and political contexts that produce human insecurity (Newman 2001), questioning existing power relations (Grayson 2004); the claim that human security tends to, and is produced through, individual empowerment (Tadjbakhsh and Chenoy 2007) to the potential for human security to give voice to the voiceless and powerless (MacFarlane 2004; Suhrke 1999). The goal, common to many humanitarian NGOs is as we will see, to place people at the center of attention by representing them as threatened victims of human insecurity and to make peoples" voices heard. Thus, it falls within this logic.

However, as many authors have argued over the last two decades, the emancipatory power of security (or securitization) is far from unproblematic. In this sense, to understand the implications of the human security framework, I find the Copenhagen School's perspective more useful. As Williams has pointed out, "in most cases, securitization is something to be avoided. While casting an issue as one of "security" may help elevate its position on the political agenda, it also risks placing that issue within the logic of threat and decision, and potentially within the contrast of friend and enemy. *Security*, accordingly, is something to be invoked with great care and, in general, minimized rather than expanded – a movement that should be sought in the name of stability, tolerance, and political negotiation, not in opposition to it" (2003, 523). Subsequently Browning and McDonald (2011) have shown how this approach not only neglects the potential negative consequences of depicting an issue within the logic of security, but, that by framing security into an emancipatory process it also fails to explore whether emancipation would be better served by other languages (such as justice and economics).

Reflecting on the link between security and emancipation, Aradau (2004a) has explained how the logic is eventually counterproductive since it reproduces the same dynamics of inclusion/exclusion and passivity that was initially challenged.

Even acknowledging the important mobilization effect of securitization for emancipation implies a logic of exceptionalism that is at odds with the democratic project. In fact, the equation of emancipation with security, legitimizes the security strategy and "endorse(s) the exclusionary logic of security and the politics that is instituted by doing security" (Aradau 2004a, 398). Moreover, her observation that NGOs, by representing a specific group (in Aradau's case study, trafficked women) as threatened victims, accomplish their scope at the expenses of those whose voices remain silent and excluded (i.e. prostitutes and asylum seekers in Aradau's example) it is quite relevant when one reflects over humanitarian communication. NGOs, along with their efforts to ensure that the (innocent) humanitarian subjects' voices are heard, are at the same time at risk of reproducing and reinforcing an exclusionary logic of security toward the dangerous 'other'. Emancipation thus can only be achieved overcoming an inclusion/exclusion logic. According to Aradau, there is a dual problem. On the one side, drawing on Balibar's writings, she notices how nobody could be emancipated by an external actor. On the other, the power of emancipation is only available for those who are considered to be members of the community and refugees and migrants are not recognized as such. Specifically, addressing the limits of human security with regards to its emancipatory potential, McCormack (2008) has pointed out how this framework not only fails to disrupt the existing power inequalities but, on the contrary, actually reproduces them. She has pointed out how the contradiction of the empowering and emancipatory promise of human security lies in its relocation of sovereignty in international community's institutions and powerful states instead of citizenship. Indeed, "there is no 'golden age' of human security policy in which the 'voiceless' and most vulnerable set international and national security policy priorities. Even before the 'war on terror', the human security agenda was set according to Western priorities—albeit in the name of the poor and disempowered." (McCormack 2008, 120). In this context, she argues that the external actors who intervene to protect human security do not have any obligation of accountability. Instead of empowering individuals, the human security framework endows international institutions and NGOs with agency.

2.4.3   Securitization and Humanitarianism

After having outlined the notion of humanitarianism and the concept of securitization – as we have seen intertwined with that of human security – it is now possible to look at the interconnections of these two dimensions. Despite humanitarianism' aspirations for neutral, impartial and independent action,[6] a large swathe of scholarship has already shown the interplay of humanitarianism with military action and security (De Lauri 2018; Donini et al. 2004, 2008; Duffield 2001; Macrae 2002; Tirman 2003). This growing literature has been mainly concerned with the

---

[6]The principles of neutrality, impartiality and independence, along with other four principles, are defined in the *Statutes of the International Red Cross and Red Crescent Movement*, adopted by the 25th International Conference of the Red Cross, Geneva, 1986.

increasingly blurred lines between humanitarian operations and military and security interventions, and their crucial political implications. Emerging scholarship has put the emphasis on a different interplay: that between humanitarianism and securitization. In her compelling analysis of trafficked women identified as illegal migrants and victims, Claudia Aradau (2004b) has brilliantly shown how humanitarianism and securitization discourses are not mutually exclusive but, rather, reciprocally constitutive. As she points out, it is not just a question of human rights framed within the 'new' concept of human security, but, rather, how two seemingly conflicting "discursive regimes are entwined and feed upon each other" (Aradau 2004b, 252). In her view, the two articulations should be considered as two interconnected systems of Foucauldian governmentality. Her perspective allows us to go beyond a dichotomic understanding of humanitarianism and securitization that, in trying to unpack how such supposedly antithetical realm could interact, misses the point of how inherently similar their premises are. In this sense, Aradau's conclusion provides a fundamental assumption underpinning this book.

Focusing on the study on the case of Syrian refugee crisis, I build on the scholarship that has shed light on the relationship between humanitarianism, security and governance in the context of refugee crises. There are a variety of studies that have investigated how humanitarian practices interrelate with security and securitization discourses. For example, Pupavac (2001a, 2005) has explored the convergence of humanitarian aid with global therapeutic governance goals. Reid (Reid 2010) has analysed the biopoliticization of humanitarianism – through practices of interventions that, instead of saving bare lives, secure them by constituting biohuman life in a political way. Rozakou (2012) has explored "the biopolitical connotations of the production of the asylum seeker– refugee as a guest" (Rozakou 2012, 563). Vaughan-Williams (2015) has investigated the zoopolitical space of humanitarian border security. Andersson (2014) has shown how the securitarian and humanitarian dimensions are part of a complex dynamic of threat and vulnerability that enables a security-humanitarian response. More recently, Gabiam (2016) has highlighted how humanitarian focus on resilience, sustainability and self-reliance to tackle the "adverse socio-economic effects"[7] of the Syria refugee crisis have been linked with global security concerns. With reference to the recent Syrian displacement to Europe, Chouliaraki (2017) has argued how security and humanitarian responses collaborate in a new moral order that she defines of hospitality, whereby the border is reaffirmed as a space of power and exclusion but simultaneously enables 'micro-connections of solidarity' that reinforce it and challenge it at the same time. Focusing on SAR operations in the Mediterranean, Cuttitta has argued that "the humanitarianization of migration and border management converges with its securitization" (Cuttitta 2017, 5).

These theories are crucial to understanding the process and practices through which humanitarian work contributes to the shaping of a security discourse and how

---

[7] United Nations Development Programme and UNHCR, *Regional Refugee and Resilience Plan 2015–2016*, 28 July 2015, accessed 15 December 2015, cited in Gabiam (2016).

securitization and humanitarianism interact in multiple ways. However, none of these studies has focused on visual representation, rather than humanitarian practices. This book intends to fill this gap. By drawing on this literature, it intends to push forward this approach to investigate the role of transnational humanitarian NGOs in global governance through the lenses of securitization theory by studying relief organizations' visual production.

Before proceeding to the next section that will theoretically introduce the NGOs communication space and the humanitarian representation of refugees, an important consideration on the 'acceptance' of the humanitarian securitization process is in order. As mentioned above, the Copenhagen School outlines a substantial difference between a securitization *move* and a securitization *process* and the difference between the two mechanisms lies precisely in the question of acceptance. Basically, in order to be able to speak of a securitization process, it is necessary that an actor's utterance in terms of security is accepted by the public. Without the acceptance, we can only speak of a securitization move. Some scholars have criticized the Copenhagen School for having overlooked the question of audience and having left the concept underdeveloped. Indeed, for some authors, the role of the audience would be so important that they have not only argued for paying more attention to the concept, but they have also proposed a refinement of the notion. For Balzacq for instance, the audience is a core element of securitization. In order to achieve its perlocutionary effect – to convince the public – the actor has in fact to "tune his/her language to the audience's experience" (Balzacq 2005, 184). In this sense, according to the author, the context, as well as the psycho-social disposition of the audience, would be crucial for the securitizing process.

Moreover, Balzacq and other scholars have pointed out that the audience should be understood as multiple audiences following distinct logics of persuasion (Balzacq 2005; Salter 2008; Vuori 2008). Another issue lies in the intended intersubjective character of the securitization process and what this would entail for the role of audience. Côté has pointed out that the audience should be understood as an active subject, "capable of having an independent effect on securitization outcomes" (Côté 2016, 543).

From all this it is quite evident that the question of audience is anything but a small detail when talking about humanitarian securitization. Indeed, this book will show that transnational humanitarian NGOs can be considered securitizing actors exactly because their discourse has been accepted. There is, therefore, a crucial node to disentangle here. On one hand, as will be discussed in the next chapter, this book will mainly focus on two 'sites' of the images[8]: production and image *per se*. It will mostly overlook the site of audience to which the literature (mainly in communication studies, sociology and psychology) has dedicated a great deal of

---

[8] With the concept of sites of images, I refer to what Rose has indicated as the different levels at which visual artefacts produce meanings: "'interpretations of visual images broadly concur that there are three sites at which the meanings of an image are made: the site(s) of the production of an image, the site of the image itself, and the site(s) where it is seen by various audiences" (Rose 2001, 16).

attention. From an IR perspective, this book, rather than focusing on the individuals' perception and reception images, is more interested in unpacking the various meanings that are made possible by the polysemic value of visual artefacts and, most importantly, their impact on the role that relief organizations play in global governance. In this sense, a certain degree of acceptance in a humanitarian securitized discourse is taken for granted not only for the "expert authority" for which they are recognized (Barnett and Duvall 2004, 171), but also implied in the role that humanitarian actors are granted – by public opinion but also state-based politics – in the management of human suffering and displacement. In other words, the fact that relief organizations are commonly considered legitimate actors to address human insecurity entails a certain degree of acceptance of their discourse. I will leave the discussion regarding the different forms that the humanitarian securitization process can take for the analytical Chaps. 4, 5 and 7. Here, I would like to conclude by acknowledging lessons derived from the literature outlined above. It has highlighted the importance of context and the existence of multiple audiences. These two crucial points will be discussed in detail in the chapter dedicated to the visual approach.

2.5   Humanitarian Communication

2.5.1   Humanitarian NGOs and Marketing

Present-day humanitarianism "has been affected by the forces of production; by the ascendance of a business discourse of markets, efficiency, accountability, and effectiveness; by shifting ideologies regarding the state's role in the economy and society; and by new funding patterns" (Barnett and Weiss 2008, 28). Following this reasoning, Stephen Hopgood (2008) provocatively asked whether Wal-Mart – one of the world's biggest multinational retail corporations – could be considered a humanitarian organization. He has highlighted the connection between contemporary humanitarian enterprise and neoliberalism, showing how the logic of business has permeated the former. It should come as no surprise, he argues, that relief organizations such as CARE, Oxfam and Save the Children have developed tight relationships with private capital, as they have to attract both public and private funding in an increasingly competitive environment. In a context in which several organizations have embraced this neoliberal logic, "humanitarian aid" has become synonymous with "aid industry", whereby NGOs become an effective donors' tool to transfer development funds to low-income countries or a way to privatize foreign assistance in a way that reduces accountability both to governments and beneficiaries (Polman 2010; Rieff 2002). As John Quelch and Nathalie Laidler-Kylander have very effectively explained, "mission effect is the surrogate for profits. Mission aligns the organization with its stakeholders, sets the boundaries for the organization, and provides the foundation on which trust is developed. Strong NGO brands succinctly articulate their missions in terms of what, how, and for whom; these missions are equivalent in many ways to brand-positioning statements" (Quelch and Laidler-Kylander 2005, 10).

Against this background, branding becomes an essential element to raise money in a competitive market. Since fundraising is fundamental to sustain NGOs' operations, branding also becomes an important part of the organizations' mission. Cottle and Nolan have highlighted how in a highly competitive environment humanitarian NGOs do not only compete for funds, but also for public attention, and have today thus assimilated a"media logic". Their relationship with the media is "at once indispensable and inimical to NGO aims and ideas of global humanitarianism" (Cottle and Nolan 2007, 763). In order to raise awareness of forgotten crises and mobilize support, aid agencies need to successfully compete within the media environment. However, by focusing on organizational branding, designing stories in a certain way and using celebrities in order to appeal to media interests, regionalizing and personalizing the assistance via downplaying the work of local organizations, and by dedicating time and resources to protect their reputation and credibility, they end up detracting from their original goal and humanitarian principles (Cottle and Nolan 2007). "On the television, in the newspaper, in the mail—the public face of the aid industry is never far away. This is a face which is quite literally a face, that of the hungry child, helpless mother, homeless refugee. Through these faces, aid agencies sell themselves and their missions; they use marketing techniques honed over the decades by businesses and non-profits" (Kennedy 2009, 1).

2.5.2 Humanitarian Communication

Over the last three decades, NGOs have used communication technologies to create an international space of solidarity and activism. This is the reason a large swathe of scholarship has identified as the precursors of an emerging civil society (DeChaine 2005). In this sense, NGOs have contributed to the symbolical and material shaping of an international community able to overcome the challenges of international solidarity (e.g., political fragmentation, economic colonization, and cultural homogenization) through their humanitarian discourses. At the same time, they have contributed to the mainstream discursive process of construction of social reality and public morality of the global civil society.

Present day humanitarian communication can be best understood in terms of mediatized humanitarianism, as Vesteergards has suggested. Mediatization is intended as the: "process whereby society to an increasing degree is submitted to, or becomes dependent on, the media and their logic resulting in enduring changes to the character, function and structure of social institutions and cultural processes. Institutions to an increasing degree become dependent on resources that the media control, and will have to submit to the rules the media operate by in order to gain access to those resources" (Vestergaard 2011, 99).

However, this is only one side of the coin for at the same time another process occurs: that of "mediation" by which social and cultural transformation in turn, affect the forms in which media can be produced and perceived. In this sense, mediatization is best understood as "a thoroughly dialectic process of mutual constitution between humanitarian organizations and audiences which themselves continually undergo transformations without homogenous or isolatable causality" (Vestergaard 2011, 100). Her perspective is particularly relevant as it underlines a characteristic of humanitarian communication that is generally overlooked: the fact that over the last decades it has been increasingly influenced and transformed by the media logic, but that at the same time, social and cultural conditions have affected the way humanitarian organizations have represented themselves and their work. While investigating the role that relief organizations play in global governance and securitization, the features of humanitarian communication, are an important dimension to address for their role in consolidating and disseminating humanitarian discourse.

The literature focusing on humanitarian communication has highlighted how it has been traditionally focused on suffering. In a famous book, Boltanski (1999) has analysed the representation of distant suffering and the moral and political implication for the spectator. For the sociologist, present day humanitarianism is undergoing a "crisis of pity" that can be read as a crisis of a theatrical conception of politics, through which the call for action in the name of humanity conceals the complicity of humanitarianism with power behind an emotionally powerful representation of suffering (Chouliaraki 2013). Boltanksy was more interested in the exploration of the rhetorical "topics" available for the expression of the spectator reaction to the spectacle of suffering. This he identified in the topics of denunciation, of sentiment and the aesthetic topic. Other authors have focused on the process of production and dissemination of suffering. Kennedy (2009) has explained how humanitarian agencies use the representation of suffering as a mean to "bridge the distance (…) in relation to a wider theoretical literature on proximity and assistance that maintains that people are less likely to respond to aid victims who are far away" (Kennedy 2009, 1).

The result of this marketing strategy, he maintained, is that of commodifying suffering with ambivalent effects. While it is functional to the organization's goal of attracting funds and raising awareness, it simultaneously contributes to the construction of the humanitarian narrative in terms of hopeless and passive victimhood.

Following this line of reasoning, but also keeping account of the changes that humanitarian communication has undergone throughout the years, Lilie Chouliaraki has introduced the useful concept of "post-humanitarianism". Reflecting both on the mediatization process, but also on the dominant representation of suffering that distinguished emotion-based humanitarian communication at least until the 1990s, she has pointed out how present-day humanitarian rhetoric has shifted to a post-emotional style (Chouliaraki 2010). She has shown how the communication of contemporary humanitarian organizations has abandoned its claims of universal morality for styles that are based on corporate branding and that rely on aesthetic topics of contemplation to reflect upon, rather than grand emotions to inspire immediate action. Post-humanitarian communication is based on puns, extremely high aesthetic quality as well as: "low intensity emotional regimes and a technological imagination of instant gratification and no justification. While still depending on realistic imagery (of the poor, the wounded or the about-to-die), the key feature of post-humanitarianism lies precisely in loosening up this 'necessary' link between seeing suffering and feeling for the sufferer, and in de-coupling emotion for the sufferer from acting on the cause of suffering. Central to the post-humanitarian sensibility is the particularization of the cause, whereby the representation of suffering becomes disembedded from discourses of morality and relies on each spectator's personal judgement of the cause for action" (Chouliaraki 2010, 119). In this new form of communication, the target of the campaign is the single individual and her/ his personalized reaction of ludic engagement, rather than the universalized and collective action that was the goal of traditional rhetoric. As Chouliaraki has observed, the problem of this new style is that it tends to reproduce a political culture of Western narcissism, a kind of sensibility that makes the emotions felt by the individual the measure of the understanding of the suffering of others.

Since the end of the nineteenth century images of human misery have been used to draw attention to forgotten humanitarian crises and create a "humanitarian imaginary" of human suffering in worlds distant for the European and American public (Fehrenbach and Rodogno 2015; Massari 2020b) in order to mobilize support and fundraise. Indeed, humanitarian communication has always utilized visual material and technologies have progressively and increasingly been introduced: photography, videos, web-based interfaces, and now the use of interactive media. It has been acknowledged for quite some time now the that photographic accounts of the Biafra famine in 1968 and subsequent humanitarian crises in the Horn of Africa and the Sahel were crucial in drawing the general public's attention to places that, to an Western audience, were and felt even more distant than today (Sontag 2003). However, throughout the years, visual humanitarian communication has been criticized for various reasons. Vivid and shocking images of suffering have been denounced for reproducing a colonial perspective. This further distances the observer and the victim (Hall 2001), contributes to the humanitarian reduction of the victim (Fassin 2007), and ultimately dehumanizes the sufferer (Benthall 1993; Lissner 1977). Representations of suffering, a characteristic trait of humanitarian communication, have also been criticized for their inherent commodification of suffering (Kennedy 2009), compassion fatigue (Moeller 1999), and their role in concealing the root political causes of humanitarian disasters (Campbell 2012).

Visual humanitarian communication has also been criticized for the use of positive images that eschew accuracy and obscure the misery of suffering (Lidchi 1999). Chouliaraki (2010) has correctly traced the connection between the use of a positive imaginary with the more recent way NGOs portray agency and the dignity of people represented. During the 1990s several organizations' communication guidelines and policy documents started to reflect the new paradigm. For example, Save the Children's *Focus on Images. The Save the Children Fund Images Guidelines* (Save the Children quoted in Manzo 2008), CARE's *Brands Standards* (CARE quoted in Kennedy 2009) and the European NGO Confederation for Relief and Development's *Code of Conduct on Images and Messages* updated in 2006 (Concord 2006) all point toward this new direction, that of avoiding undignifying representations of passive victimhood. This new approach has proceeded hand-in-hand with new humanitarianism's shift of paradigm from attention to people in need to depiction of people as rights holders. Over the years, humanitarian communication has tended to move away from images of starving babies to depictions of displaced children deprived by conflict of their right to education. Similarly, pictures of indistinguishable faces of people in distress have gradually given way to images portraying resilient individuals with very distinct individual features. What all these 'new' images, focusing on positive representations, have in common is the depiction of refugees as people needing to be protected from serious violations of their human rights. This book will argue that both positive and negative representations of humanitarian subjects end up passivizing the people represented and reducing them to an essentialized character of 'victim.' This occurs even more often in the representation of refugees.

2.5.2.1 Visual Representation of Refugees

In the field of visual humanitarian communication important scholarly work has focused on the representation of refugees. In a seminal article on the topic, Malkki (1996) pointed out how photographic accounts of displacement have consistently portrayed refugees as a "sea of humanity". "No names, no funny faces, no distinguishing marks, no esoteric details of personal style enter, as a rule, into the frame of pictures of refugees when they are being imagined as a sea of humanity" (Malkki 1996, 388). These representations of people as universal victims contributed to the dehistoricization and depoliticization of the refugee experience in favor of the construction of the 'universal humanitarian subject'. Depictions of this kind imply the silencing of the people finding themselves classified in a "refugee category", while legitimizing the professional humanitarian claim of producing authoritative accounts on refugees.

Even though, as Malkki has argued, representations of raw and indistinguishable humanity dominated the media visual narrative, depictions in terms of helplessness and speechlessness have been also consistently typical in conventionalized photos of refugees: those with women and children, whose corporality was only embodying their role of universal passive victims. More recently, Bleiker et al. (2013) have confirmed this trend, studying how Australian media represents refugees. The analysis showed that images of medium or large groups of people have prevailed in the visual depiction of people on the move, while pictures of individuals with distinguishing traits – commonly considered more likely to provoke feelings of empathy and compassion (Bleiker et al. 2014) – have remained relatively absent. The effect of this dehumanizing visual framing, they argue, has been that of reinforcing an image of refugees associated with threat and security concerns, rather than a humanitarian challenge. However, even when the focus of the representation is the individual and his/her distinguishing traits, humanitarian narrative tends to depict her/him as inexorably belonging to a universal refugeeness (Nyers 1999). It is not a coincidence, Nyers argues, that in a publication titled "What is like to be a refugee?"

the cover photo portrays a shirt hanging outside of a shelter with no human bodies or faces around. The object is there to represent the universal situation of the humanitarian subject and the relative feeling of loss and emptiness. This invisibility of the individual persists also when single people are portrayed to represent the experience or the loss of rights of a whole category.

The issues of speechlessness and invisibility are particularly important if considered in relation to present day NGOs' efforts to portray people with agency and empowerment. Harrel-Bond quotes a refugee intervening in an Oxford conference on *Assistance to Refugees: Alternative Viewpoints*: "Why not publicize our energy and our power to help ourselves? [...] We talk about UNHCR and we talk about NGOs, but we forget the refugees themselves. We forget the power they have to help themselves" (Harrell-Bond 1985, 4). According to her, this is not possible, because it would undermine the very legitimacy of humanitarian organizations work with refugees. For Harrell-Bond "humanitarian agencies are in a straitjacket with little else than human misery upon which to base their appeals" (Harrell-Bond 1985, 4). Although trenchantly expressed, this is a point that needs to be taken into account when discussing positive images and how NGOs attempt to show people with agency and empowerment. In fact, such kind of representation could have the counter effect of completely erasing the indications of need for mobilization and intervention, which is one of the primary goals of humanitarian communication. Moreover, even when the objective of humanitarian depiction is that of giving voice, individuality, and empowerment to refugees, the outcomes are often similar to those identified by Malkki in her analysis of traditional representations of refugees. As Rajaram has observed, refugees continue to be denied the possibility to produce political narratives while the account of their experience remains a prerogative of Western relief agencies, through which "refugee lives become a site where Western ways of knowing are reproduced" (Rajaram 2002, 247).

This is particularly pertinent as the ways refugees are represented, and therefore come to be known to the general public, are crucial in creating the "conditions of possibility for cultures of hospitality" (Bleiker et al. 2014, 192). Bleiker et al. have observed how Cold War images of refugees privileged portraits of Eastern bloc citizens escaping to the West where they were celebrated as heroes surviving political oppression. Very differently, according to the authors, stereotypical images of today's refugees are dominated by overcrowded boats or portraits of passive women and children in need of external intervention. The consequence of this visual narrative is the reproduction and reinforcement of security discourses on displacement and a binary understanding of people on the move either as passive victims in need of protection or as threat to the hosting communities. Beside the influence of images of cultures of hospitality, visual representations have also been studied for their potential to affect social and political change. In this sense, shocking images such as the sadly famous one of Alan Kurdi washed ashore, have been found to produce strong feelings relating to identity, emotions and beliefs and social media engagement right after the publication of the picture (Thomas et al. 2018). However, the dramatic photo did not produce a radical change in discourses and representations of refugees (Bozdag and Smets 2017; Slovic et al. 2017), let alone changes in migration regimes (Burns 2015; The Independent 2016, 2017). On the contrary, as Achilli (2019) has observed, its long-term impact has mostly contributed to the demonization of irregular migration and shifted attention from States' responsibility to help those in need toward the smugglers who exploit them.

2.6   To Be Continued…

The present chapter has outlined the conceptual framework that underpins the research. In doing so it has addressed the theoretical debate on the interconnections between humanitarianism and politics to contextualize the analysis and the assumption on which it is based. It has successively focused on the theoretical presentation of the role that civil society and NGOs in particular play in global governance to then specifically address the role of transnational humanitarian organizations in the international political arena. The third section has introduced the securitization theory, as conceptualized by the Copenhagen School, and outlined how this approach is particularly suitable for the purpose of this research. The last part has been devoted to the presentation of the humanitarian communication context and the specificities and characteristic of humanitarian communication, along with review of the relevant literature on the humanitarian representation of refugees. However, given the particular attention that this research accords to visuality, this chapter cannot be concluded without addressing the conceptual framework of a visual approach. The next chapter will focus on the visual and its theoretical implications.

References

Abiew, F. K. (2003). NGO-military relations in peace operations. *International Peacekeeping, 10*(1), 24–39.

Achilli, L. (2019). Waiting for the smuggler: Tales across the border. *Public Anthropologist, 1*(2), 194–207.

Andersson, R. (2014). *Illegality, Inc.: Clandestine migration and the business of bordering Europe.* Oakland: University of California Press.

Anheier, H. K., Glasius, M., & Kaldor, M. (2001). *Global civil society.* Oxford: Oxford University Press.

Aradau, C. (2004a). Security and the democratic scene: Desecuritization and emancipation. *Journal of International Relations and Development, 7*(4), 388–413.

Aradau, C. (2004b). The perverse politics of four-letter words: Risk and pity in the securitisation of human trafficking. *Millennium, 33*(2), 251–277.

Balzacq, T. (2005). The three faces of securitization: Political agency, audience and context. *European Journal of International Relations, 11*(2), 171–201.

Barnett. (2013). Humanitarian governance. *Annual Review of Political Science, 16*, 379–398.

Barnett, L. (2002). Global governance and the evolution of the international refugee regime. *International Journal of Refugee Law, 14*(2_and_3), 238–262.

Barnett, M. (2009). Evolution without progress? Humanitarianism in a world of hurt. *International Organization, 63*(4), 621–663.

Barnett, M. (2011). *Empire of humanity: A history of humanitarianism*. Ithaca: Cornell University Press.

Barnett, M., & Duvall, R. (2004). *Power in global governance*. Cambridge: Cambridge University Press.

Barnett, M., & Snyder, J. (2008). The grand strategies of humanitarianism. In M. Barnett & T. G. Weiss (Eds.), *Humanitarianism in question. Politics, power, ethics*. Ithaca: Cornell University Press.

Barnett, M., & Weiss, T. G. (2008). *Humanitarianism in question: Politics, power, ethics*. Ithaca: Cornell University Press.

Benthall, J. (1993). *Disasters, relief and the media*. London: IB Tauris.

Bleiker, R., Campbell, D., & Hutchison, E. (2014). Visual cultures of inhospitality. *Peace Review, 26*(2), 192–200.

Bleiker, R., Campbell, D., Hutchison, E., & Nicholson, X. (2013). The visual dehumanisation of refugees. *Australian Journal of Political Science, 48*(4), 398–416.

Boltanski, L. (1999). *Distant suffering: Morality, media and politics*. Cambridge: Cambridge University Press.

Booth, K. (1991a). Security and emancipation. *Review of International Studies, 17*(4), 313–326.

Booth, K. (1991b). Security in anarchy: Utopian realism in theory and practice. *International Affairs (Royal Institute of International Affairs 1944-), 67*, 527–545.

Bozdag, C., & Smets, K. (2017). Understanding the images of Alan Kurdi with "small data": A qualitative, comparative analysis of tweets about refugees in Turkey and Flanders (Belgium). *International Journal of Communication, 11*, 4046–4069.

Brems, E. (1997). Enemies or allies? Feminism and cultural relativism as dissident voices in human rights discourse. *Human Rights Quarterly, 19*(1), 136–164.

Brown, C. (2014). *International society, global polity: An introduction to international political theory*. Los Angeles: Sage.

Browning, C. S., & McDonald, M. (2011). The future of critical security studies: Ethics and the politics of security. *European Journal of International Relations, 19*(2), 235–255.

Burns, A. (2015). Discussion and action: Political and personal responses to the Aylan Kurdi images. In *The iconic image on social media: A rapid research response to the death of Aylan Kurdi. Visual social media lab December* (pp. 38–39). Sheffield: University of Sheffield.

Buzan, B., & Hansen, L. (2009). *The evolution of international security studies*. Cambridge University Press.

Buzan, B., Wæver, O., & De Wilde, J. (1998). *Security: A new framework for analysis*. Boulder: Lynne Rienner Publishers.

Campbell, D. (2012). The iconography of famine. In *Picturing atrocity: Photography in crisis* (pp. 79–92). London: Reaktion Books.

Chandler, D. (2001). The road to military humanitarianism: How the human rights NGOs shaped a new humanitarian agenda. *Human Rights Quarterly, 23*(3), 678–700.

Chandler, D. (2002). *FROM KOSOVO TO KABUL: Human rights and international intervention*. London: Pluto Press.

Chandler, D. (2004). *Constructing global civil society: Morality and power in international relations*. Basingstoke: Springer.

Charny, J. R. (2004). Upholding humanitarian principles in an effective integrated response. *Ethics & International Affairs, 18*(2), 13–20.

Chouliaraki, L. (2010). Post-humanitarianism: Humanitarian communication beyond a politics of pity. *International Journal of Cultural Studies, 13*(2), 107–126.

Chouliaraki, L. (2013). *The ironic spectator: Solidarity in the age of post-humanitarianism*. Oxford: Wiley.

Chouliaraki, L. (2017). Symbolic bordering: The self-representation of migrants and refugees in digital news. *Popular Communication, 15*(2), 78–94.

Collinson, S., & Elhawary, S. (2012). *Humanitarian space: Trends and issues*. London: Overseas Development Institute.

Concord. (2006). *Code of conduct on images and messages*. https://concordeurope.org/wp-content/uploads/2012/09/DEEEP-BOOK-2014-113.pdf?c676e3&c676e3

Coomaraswamy, R. (1994). To bellow like a cow: Women, ethnicity, and the discourse of rights. *Human Rights of Women: National and International Perspectives, 39*, 56.

Côté, A. (2016). Agents without agency: Assessing the role of the audience in securitization theory. *Security Dialogue, 47*(6), 541–558.

Cottle, S., & Nolan, D. (2007). Global humanitarianism and the changing aid-media field: "Everyone was dying for footage". *Journalism Studies, 8*(6), 862–878.

Cuttitta, P. (2017). Repoliticization through search and rescue? Humanitarian NGOs and migration management in the central mediterranean. *Geopolitics, 23*, 1–29.

Cutts, M. (1998). Politics and humanitarianism. *Refugee Survey Quarterly, 17*(1), 1–15.

Dany, C. (2015). Politicization of humanitarian aid in the European Union. *European Foreign Affairs Review, 20*(3), 419–437.

Davies, T. R. (2014). *NGOs: A new history of transnational civil society*. New York: Oxford University Press.

De Lauri, A. (2016). *The politics of humanitarianism. Power, ideology and aid*. London: IB Tauris.

De Lauri, A. (2018). Humanitarian militarism and the production of humanity. *Social Antrhopology, 27*(1), 84–99.

DeChaine, D. R. (2005). *Global humanitarianism: NGOs and the crafting of community*. Lanham: Lexington Books.

Dijkzeul, D., & Moke, M. (2005). Public communication strategies of international humanitarian organizations. *International Review of the Red Cross, 87*(860), 673–691.

Dillon, M., & Reid, J. (2000). Global governance, liberal peace, and complex emergency. *Alternatives, 25*(1), 117–143.

Donini, A., Fast, L., Hansen, G., Harris, S., Minear, L., Mowjee, T., & Wilder, A. (2008). *The state of the humanitarian enterprise*. Boston: Feinstein International Center, Tufts University.

Donini, A., Minear, L., Azarbaijani-Moghaddam, S., Hansen, G., Mowjee, T., Purushotma, K., Smillie, I., Stites, E., & Zeebroek, X. (2006). *Humanitarian agenda 2015: Principles, power, and perceptions*. Medford: Feinstein International Center.

Donini, A., Minear, L., & Walker, P. (2004). The future of humanitarian action: Mapping the implications of Iraq and other recent crises. *Disasters, 28*(2), 190–204.

DuBois, M. (2007). Protection: The new humanitarian fig-leaf. *Dialogue, 4*, 1.

Duffield, M. (2001). *Global governance and the new wars: The merging of development and security*. London: Zed Books Ltd..

Duffield, M. (2007). *Development, security and unending war: Governing the world of peoples*. Cambridge: Polity.

Duffield, M. (2014). *Global governance and the new wars: The merging of development and security*. London: Zed Books Ltd..

Duffield, M., Macrae, J., & Curtis, D. (2001). Politics and humanitarian aid. *Disasters, 25*(4), 269–274.

Edwards, M. (2009). *Civil society*. Cambridge: Polity.

Escobar, A. (2011). *Encountering development: The making and unmaking of the third world*. Princeton: Princeton University Press.

Fassin, D. (2007). Humanitarianism as a politics of life. *Public Culture, 19*(3), 499–520.

Fehrenbach, H., & Rodogno, D. (2015). "A horrific photo of a drowned Syrian child": Humanitarian photography and NGO media strategies in historical perspective. *International Review of the Red Cross, 97*(900), 1121–1155.

Ferguson, J. (1990). *The anti-politics machine: 'Development', depoliticization and bureaucratic power in Lesotho*. Minnesota: CUP Archive.

Fiori, J. E. (2013). *The discourse of western humanitarianism*. Paris: Humanitarian Affairs Think Tank.

Florini, A. M. (2012). *The third force: The rise of transnational civil society*. Brookings Institution Press.

Foucault, M. (1991). *The foucault effect: Studies in governmentality*. Chicago: University of Chicago Press.

Fox, F. (2001). New humanitarianism: Does it provide a moral banner for the 21st century? *Disasters, 25*(4), 275–289.

Gabiam, N. (2016). Humanitarianism, development, and security in the 21st century: Lessons from the Syrian refugee crisis. *International Journal of Middle East Studies, 48*(2), 382–386.

Grayson, K. (2004). A challenge to the power over knowledge of traditional security studies. *Security Dialogue, 35*(3), 357–357.

Hall, S. (2001). The west and the rest. In S. Hall & B. Gieben (Eds.), *Formations of modernity* (pp. 257–330). Cambridge: Open University Press\Blackwell.

Hansen, G. (2007). *Taking sides or saving lives: Existential choices for the humanitarian enterprise in Iraq* (Feinstein International Center Briefing Paper).

Hansen, L. (2011). Theorizing the image for security studies: Visual securitization and the Muhammad Cartoon Crisis. *European Journal of International Relations, 17*(1), 51–74.

Harmer, A. (2008). Integrated missions: A threat to humanitarian security? *International Peacekeeping, 15*(4), 528–539.

Harrell-Bond, B. (1985). Humanitarianism in a straitjacket. *African Affairs, 84*(334), 3–13.

Heck, A., & Schlag, G. (2013). Securitizing images: The female body and the war in Afghanistan. *European Journal of International Relations, 19*(4), 891–913.

Higgins, R. (1993). The new United Nations and former Yugoslavia. *International Affairs (Royal Institute of International Affairs 1944-), 69*, 465–483.

Hilhorst, D., & Jansen, B. J. (2010). Humanitarian space as arena: A perspective on the everyday politics of aid. *Development and Change, 41*(6), 1117–1139.

Hopgood, S. (2008). Saying no to wal-mart. In M. Barnett & T. G. Weiss (Eds.), *Humanitarianism in question: Politics*. Ithaca: Cornell University Press.

Hudson, A. (2001). NGOs' transnational advocacy networks: From 'legitimacy'to "political responsibility"? *Global Networks, 1*(4), 331–352.

Huysmans, J. (1998). Revisiting Copenhagen: Or, on the creative development of a security studies agenda in Europe. *European Journal of International Relations, 4*(4), 479–505.

Kaldor, M. (1999). Transnational civil society. In T. Dunne & N. J. Wheeler (Eds.), *Human rights in global politics* (pp. 195–213). Cambridge: Cambridge University Press.

Kaldor, M. (2003). The idea of global civil society. *International Affairs, 79*(3), 583–593.

Kennedy, D. (2009). Selling the distant other: Humanitarianism and imagery—Ethical dilemmas of humanitarian action. *The Journal of Humanitarian Assistance, 28*, 1–25.

Krause, K., & Williams, M. C. (1996). Broadening the agenda of security studies: Politics and methods. *Mershon International Studies Review, 40*(Supplement_2), 229–254.

Leader, N. (2000). *The politics of principle: The principles of humanitarian action in practice* (Vol. 2). London: Humanitarian Practice Group (HPG).

Lidchi, H. (1999). Finding the right image: British development NGOs and the regulation of imagery. In *Culture and global change* (pp. 87–101). London/New York: Routledge.

Lissner, J. (1977). *The politics of altruism: A study of the political behaviour of voluntary development agencies*. Geneva: Lutheran World Federation, Department of Studies.

MacFarlane, S. N. (2004). A useful concept that risks losing its political salience. *Security Dialogue, 35*(3), 368–369.

Macrae, J. (2002). *The new humanitarianisms: A review of trends in global humanitarian action*. London: Humanitarian Policy Group, Overseas Development Institute.

Macrae, J., & Leader, N. (2001). Apples, pears and porridge: The origins and impact of the search for 'coherence' between humanitarian and political responses to chronic political emergencies. *Disasters, 25*(4), 290–307.

Malkki, L. H. (1996). Speechless emissaries: Refugees, humanitarianism, and dehistoricization. *Cultural Anthropology, 11*(3), 377–404.

Manzo, K. (2008). Imaging humanitarianism: NGO identity and the iconography of childhood. *Antipode, 40*(4), 632–657.

48 2 Humanitarianism, Securitization and Humanitarian Communication

Massari, A. (2020a). Human security. In *Humanitarianism: Keywords* (pp. 91–93). Boston: Brill.

Massari, A. (2020b). Photography. In *Humanitarianism* (pp. 158–160). Leiden: Brill. https://doi.org/10.1163/9789004431140_074.

McCormack, T. (2008). Power and agency in the human security framework. *Cambridge Review of International Affairs, 21*(1), 113–128.

McDonald, M. (2008). Securitization and the construction of security. *European Journal of International Relations, 14*(4), 563–587.

Metcalfe, V., Haysom, S., & Gordon, S. (2012). *Trends and challenges in humanitarian civil-military coordination: A review of the literature.* London: The Humanitarian Policy Group.

Moeller, S. D. (1999). *Compassion fatigue: How the media sell disease, famine, war and death.* London: Routledge/Psychology Press.

Nascimento, D. (2015, February 18). One-step forward, two steps back? Humanitarian challenges and dilemmas in crisis settings. *Journal of Humanitarian Assistance.*

Newman, E. (2001). Human security and constructivism. *International Studies Perspectives, 2*(3), 239–251.

Nyers, P. (1999). Emergency or emerging identities? Refugees and transformations in world order. *Millennium, 28*(1), 1–26.

OCHA, UN. (2003). *Glossary of humanitarian terms: In relation to the protection of civilians in armed conflict.* New York: United Nations. UN OCHA (2012): OCHA on message: Humanitarian principles. New York: UN OCHA. UN OCHA (Nedatovno): Consolidated appeal process: About the process, United Nations Office for the coordination of humanitarian affairs 15(11): 2013.

Olivius, E. (2014). *Governing refugees through gender equality: Care, control, emancipation.* Umeå: Umeå Universitet.

Pandolfi, M. (2003). Contract of mutual (in) difference: Governance and the humanitarian apparatus in contemporary Albania and Kosovo. *Indiana Journal of Global Legal Studies, 10*(1), 369–381.

Polman, L. (2010). *The crisis caravan: What's wrong with humanitarian aid?* New York: Metropolitan Books.

Pugh, M. (1998). Military intervention and humanitarian action: Trends and issues. *Disasters, 22*(4), 339–351.

Pupavac, V. (2001a). Misanthropy without borders: The international children's rights regime. *Disasters, 25*(2), 95–112.

Pupavac, V. (2001b). Therapeutic governance: Psycho-social intervention and trauma risk management. *Disasters, 25*(4), 358–372.

Pupavac, V. (2005). Human security and the rise of global therapeutic governance. *Conflict, Security & Development, 5*(2), 161–181. https://doi.org/10.1080/14678800500170076.

Pupavac, V. (2006). The politics of emergency and the demise of the developing state: Problems for humanitarian advocacy. *Development in Practice, 16*(3–4), 255–269.

Quelch, J. A., & Laidler-Kylander, N. (2005). *The new global brands: Managing non-government organizations in the 21st century.* Toronto: Southwestern Publishing.

Rajaram, P. K. (2002). Humanitarianism and representations of the refugee. *Journal of Refugee Studies, 15*(3), 247–264. https://doi.org/10.1093/jrs/15.3.247.

Reid, J. (2010). The biopoliticization of humanitarianism: From saving bare life to securing the biohuman in post-interventionary societies. *Journal of Intervention and Statebuilding, 4*(4), 391–411.

Rieff, D. (2002). *Humanitarianism in crisis* (pp. 111–121). New York: Foreign Affairs.

Rieff, D. (2003). *A bed for the night: Humanitarianism in crisis.* New York: Simon and Schuster.

Rist, G. (2014). *The history of development: From Western origins to global faith.* London: Zed Books Ltd..

Rose, G. (2001). *Visual methodologies: An introduction to researching with visual materials.* London: Sage.

Rosenau, J. N. (1995). Governance in the twenty-first century. *Global Governance, 1*(1), 13–43.

Rozakou, K. (2012). The biopolitics of hospitality in Greece: Humanitarianism and the management of refugees. *American Ethnologist, 39*(3), 562–577.

Salter, M. B. (2008). Securitization and desecuritization: A dramaturgical analysis of the Canadian Air Transport Security Authority. *Journal of International Relations and Development, 11*(4), 321–349.

Sartori, G. (1970). Concept misformation in comparative politics. *American Political Science Review, 64*(4), 1033–1053.

Sending, O. J., & Neumann, I. B. (2006). Governance to governmentality: Analyzing NGOs, states, and power. *International Studies Quarterly, 50*(3), 651–672.

Slim, H. (2003). Is humanitarianism being politicised? A reply to David Rieff. In *The Dutch Red Cross Symposium on Ethics in Aid*, The Hague, 8th October 2003, Vol. 8.

Slim, H. (2015). *Humanitarian ethics: A guide to the morality of aid in war and disaster*. New York: Oxford University Press.

Slovic, P., Västfjäll, D., Erlandsson, A., & Gregory, R. (2017). Iconic photographs and the ebb and flow of empathic response to humanitarian disasters. *Proceedings of the National Academy of Sciences, 114*(4), 640–644.

Sontag, S. (2003). *Regarding the pain of others*. London: Hamish Hamilton.

Stoddard, A. (2003). *Humanitarian NGOs: Challenges and trends* (Vol. 12). London: HUmanitarian Policy Group (HPG) Briefing. HPG.

Suhrke, A. (1999). Human security and the interests of states. *Security Dialogue, 30*(3), 265–276.

Tadjbakhsh, S., & Chenoy, A. (2007). *Human security: Concepts and implications*. London: Routledge.

Terry, F. (2013). *Condemned to repeat: The paradox of humanitarian action*. Ithaca: Cornell University Press.

The Independent. (2016). It's been two years since Alan Kurdi's death and things are much worse. *The Independent*. http://www.independent.co.uk/voices/syrian-refugees-libya-two-years-alan-kurdis-death-a7925616.html

The Independent. (2017). 8 charts that show how the refugee crisis has changed since the death of Alan Kurdi. *The Independent*. http://www.independent.co.uk/news/world/europe/alan-kurdi-death-anniversary-refugee-crisis-8-charts-how-world-has-changed-a7220741.html

Thomas, E., McGarty, C., & Smith, L. (2018). Refugee crisis: The immediate and lasting impacts of powerful images. *The Conversation*.

Tirman, J. (2003). The new humanitarianism: How military intervention became the norm. *Boston Review, 28*(6).

Vaughan-Williams, N. (2015). "We are not animals!" humanitarian border security and zoopolitical spaces in EUrope1. *Political Geography, 45*, 1–10.

Vestergaard, A. (2011). Distance and suffering: Humanitarian discourse in the age of mediatization. *Samfundslitteratur*.

Vuori, J. A. (2008). Illocutionary logic and strands of securitization: Applying the theory of securitization to the study of non-democratic political orders. *European Journal of International Relations, 14*(1), 65–99.

Wæver, O. (1995). Securitization and desecuritization. In R. Lipschutz (Ed.), *On security* (pp. 46–86). New York: Columbia University Press.

Weiss, T. G. (1999). Principles, politics, and humanitarian action. *Ethics & International Affairs, 13*, 1–22.

Weiss, T. G. (2000). Governance, good governance and global governance: Conceptual and actual challenges. *Third World Quarterly, 21*(5), 795–814.

Weiss, T. G., & Gordenker, L. (1996). *NGOs, the UN, and global governance*. Boulder: Lynne Rienner.

Williams, M. C. (2003). Words, images, enemies: Securitization and international politics. *International Studies Quarterly, 47*(4), 511–531.

Wyn Jones, R. (1999). *Security, strategy, and critical theory*. Boulder: Lynne Rienner Publishers.

Žižek, S. (2005). Against human rights. In *Wronging rights?: Philosophical challenges to human rights* (pp. 149–167). Hoboken: Taylor & Francis.

Chapter 3
A Visual Approach

3.1 Framing the Field

In the analysis of humanitarian discourse(s), I use 'discourse' in a Foucauldian sense as a system of representation of knowledge and meanings situated in a particular time and space (Foucault 1971, 1972, 1980). According to the philosopher, the concept of discourse is strictly interrelated with the production of truth and relations of power: "What I mean is this: in a society such as ours, but basically in any society, there are manifold relations of power which permeate, characterise and constitute the social body, and these relations of power cannot themselves be established, consolidated nor implemented without the production, accumulation, circulation and functioning of a discourse. There can be no possible exercise of power without a certain economy of discourses of truth which operates through and on the basis of this association. We are subjected to the production of truth through power and we cannot exercise power except through the production of truth" (Foucault 1980, 93).

Critical Discourse Analysis (CDA) is the perfect starting point for framing the theoretical field of my methodological approach. Building on the Critical Linguistic scholarship that since the 1970s has been concerned with the relationship between language and power (Blommaert and Bulcaen 2000), CDA is "fundamentally concerned with analysing opaque as well as transparent structural relationships of dominance, discrimination, power and control as manifested in language. In other words, CDA aims to investigate critically social inequality as it is expressed, signalled, constituted, legitimized and so on by language use (or in discourse)" (Wodak and Meyer 2009, 2).

Three main approaches have dominated CDA research. The first, elaborated by Fairclough (1992), considers language as discursive practice. The second (Wodak 2001) has put the emphasis on the historical dimension, while van Dijk (2015) has focused on the social cognitive aspect of discourse. What the three approaches share is a critical perspective that differentiates CDA from classical discourse analysis.

© The Author(s) 2021
A. Massari, *Visual Securitization*, IMISCOE Research Series,
https://doi.org/10.1007/978-3-030-71143-6_3

The locus of critique has to be found in the problematization of power relations and the impact of ideology on discourse patterns (Blommaert and Bulcaen 2000).

Since its origin, CDA has primarily looked at discourse through the lenses of text, overlooking other modalities of expression and particularly the visual dimension (Wang 2014). Starting from the mid-1990s, a growing group of scholars (Slembrouck et al. 1995; Kress and Van Leeuwen 1996; Rose 2001) have stressed the importance of including visual material in analysis and started focusing on visual methodologies. The importance conferred to the visual dimension in academic research became crucial not only because of the massive presence of images of all kinds (such as photography, television, art, or advertisements) in our contemporary visual landscape, but also for the acknowledgement of the pivotal role of visuality in the process of meaning production and exchange, particularly in the Western society (Rose 2001). The term 'visuality' refers to the "ways we see, how we are able, allowed, or made to see, and how we see this seeing and the unseeing therein" (Foster 1988, ix). Since the world can be seen in different ways and the different way of seeing have different social impacts, the analysis of images becomes crucial to grasp the effects of hegemonic visualities in reinforcing dynamics of power and social difference (Haraway 1991).

With the same interest in the question of visual representation, and a specific focus on International Political Theory, Roland Bleiker (2001) has contributed to the debate with a seminal article on the *Aesthetic Turn*. Starting from the observation of the increasingly wider diffusion of images representing international political events, and "their highly arbitrary nature" (Bleiker 2001, 509), the author emphasised the importance of locating politics in the differences between what is being represented and its representation. Following Jacques Derrida (1967), this approach sees the representation as an interpretation of the truth. Therefore, a political event should never be investigated *per se*, but its representation should, rather, be at the centre of the analysis so as to unveil the "sets of true statements" beyond it (Bleiker 2001, 512). In fact, argues the author, although the human tendency is to trust the resemblance of what is represented with reality – part of the human "desire to order the world" (Bleiker 2001, 515) – we should acknowledge that representation is power.

Over the last decades, several authors have focused on visuality in International Politics (see among others Robinson 1999; Boltanski 1999; Shapiro 1999; Bleiker and Kay 2007; Campbell 2007). Particularly, an emerging body of literature of IR and security studies has highlighted the relationship between visuality and security, focusing on different topics, including the political implications of representations (Campbell 2003); cartography (Shapiro 2007); the politics of security and surveillance (Andersen and Möller 2013); borders (Andersson 2012); political cartoons (Hansen 2011); science fiction (Weldes 2006); images of (post) 9/11 (Möller 2007; Weber 2006), and iconology (Heck and Schlag 2013).

Among these authors, Lene Hansen on one side and Heck and Schlag on the other, have also offered some theoretical insights for the specific study of visuality and securitization that are particularly relevant for this book. Drawing on Buzan's concept of securitization, Hansen has proposed an "intertextual framework" (Hansen

2011, 55) for the study of visual securitization which is able to investigate the ways in which visuality interrelates with other images (inter-visuality) and with words (intertextuality). According to the author, the intertextual framework is fundamental in order to explore the role of images in creating or participating in security discourse. She proposes four components for analysis: the image *per se*, the immediate intertext, the larger policy discourse, and the textual element. Hansen's model is based on the specificity of images and the distinctive way they securitize an issue. Not only is it important to consider the particular features of images (such as immediacy, ambiguity and circulability), but also the various strategies of security depiction and the different genres of visual representations (including cartoons and other drawings, photography, and video). There are three aspects in Hansen's approach to visual securitization that I find particularly relevant for the purpose of this study. Firstly, the implication of the circulability of images, that makes it possible to envisage the existence of non-elite securitizing actors. Second, the emphasis placed on the diverse "epistemic-political claims" (Hansen 2011, 53) of the different visual genres, that help in the problematization of photography in particular. Last, but not least, is her attention to inter-visuality, inter-textuality and the wider policy discourse as fundamental elements to contextualize the different meaning of images in time and space.

Drawing upon Hansen' seminal article, this study seeks to expand this method. Not only will it investigate images, but also include in the analysis photo captions, interviews with image producers, NGOs communication strategies, and relief organizations' humanitarian and advocacy positions. There are two minor, yet substantial, aspects in which my study differs from Hansen's framework. The first relates to the methodological tools utilized to carry out the analysis of the images – and specifically my selection of a combination of visual social semiotics and iconology methodologies. The second is a more theoretical point. In her understanding, images are understood as having a limited securitizing potential. As they are unable to speak for themselves, images always need an actor – able to speak – to activate their securitization potential. I intend, instead, to explore humanitarian NGOs' photographic accounts of Syrian displaced people assuming that images have an intrinsic securitizing potential. In this sense, the approach proposed by Heck and Schlag (2013) – looking at securitization through iconography – seems very useful to complement my analysis.

By focusing on the performativity power of visuality, Heck and Schlag (2013) draw on the iconological approach to theorize "the image as an iconic act understood as an act of showing and seeing" (Heck and Schlag 2013, 891). According to their method, images should be interpreted with their social context in mind, as images *per se*. The method proposed by Heck and Schlag (2013) to unveil securitization processes is based on three stages. They describe these as: "the pre-iconic description'", "the iconographic analysis", and the "iconological interpretation" of visual representation. The attention Heck and Schlag give to the potentiality of images to securitize through myth creation and narratives of justification are particularly apposite when investigating the book's main question and unpacking humanitarian discourse(s) on the Syrian refugee crisis and NGOs' role within global governance and global security.

3.2   A Semiotic Analysis of Images

The considerations and the different approaches outlined above are important to situate the analysis within a theoretical framework that considers that images and their study can unveil different humanitarian narratives, have securitization potential and drive the dynamics of constitution and dissemination of humanitarian discourses. However, the disentanglement of these mechanisms of knowledge production, and the power of the humanitarian discourse, require a certain level of operationalization. Semiotics is the perfect starting point to introduce the methodology selected for this study.

Semiotics is an area of research interested in the study of signs. With its origin in the ancient Greek world, semiotics is today applied in a wide range of different disciplines such as linguistics, religious studies, media and cartography (Nöth 2011). In semiotics, the sign (either an imagined or material sign) has to be understood in relation to both its referent object and the mental image or idea evoked (Peirce 1931, vol. 2). Its visual declination, visual semiotics, emerged in the 1960s with specific attention to visual language.

According to one of its founding fathers, Roland Barthes, there are two levels of meaning that need to be addressed in the semiotic analysis of images: denotation and connotation (Barthes 1972). The first step of analysis focuses on the identification of what van Leeuwen calls "literal message" (Van Leeuwen 2001, 94) – the Barthian "denotation" – and answers the question of what is depicted in the image. The second analytic stage is connotation and refers to ideas, values and concepts that are represented in the image. This level of analysis aims at identifying the cultural interpretations linked to specific aspect of images He argues that "such connotative meanings – in Mythologies (1972) Barthes called them 'myths' – are first of all very broad and diffuse concepts which condense everything associated with the represented people, places or things into a single entity (…). Secondly, they are ideological meanings, serving to legitimate the status quo and the interests of those whose power is invested in it" (Van Leeuwen 2001, 97).

Despite visual semiotics' crucial importance in answering questions related to what is represented in the image and what are the meanings of the representation, there are two aspects in Barthes' perspective that limit the potentiality of the analysis (Van Leeuwen 2001). The first has to do with the non-problematization of the concepts of denotation and connotation. Barthes considers the first level of meaning as if what is represented corresponds to reality without the interference of any encoding mechanism, without ambiguity, or without the possibility of different interpretations. Something similar happens with regards to the concept of connotation. The problem with this term is that, although its exploration is able to shed light on the process of condensation of values associated with the subject in a single image (and at the same legitimizing its representation), it considers the underling meaning as universally understood by different people in different times and places. These shortcomings result in a narrow focus on visual semiotics for the visual text, the lexis of the image, and an overlooking of the context, the visual syntax. In a visual analysis which takes into account intertextuality and the importance of the wider discourse around the images, the attention to the context is, on the contrary, crucial.

In this sense, social semiotics, with its emphasis on social dimensions, seems more able to grasp the social implications of visual material. In fact, this discipline is concerned with "the social dimensions of meaning in any media of communication, its production, interpretation and circulation, and its implications in social processes, as cause or effect" (Semiotics Encyclopedia Online 2018). With a particular attention to the study of images in their social context, visual social semiotics adds two additional levels to the representational level of analysis that I have outlined above: the interactional and the compositional. The first refers to the way what is represented interacts with the viewer. The second is concerned with the way images are included in the wider visual syntax.

In an article devoted to social semiotics in visual communication, Carey Jewitt and Rumiko Oyama (2001), situate the main difference between the structuralist school of semiotics and social semiotics in the notion of "semiotic resources". The authors define resources as "at once the products of cultural histories and the cognitive resources we use to create meaning in the production and interpretation of visual and other messages" (Jewitt and Oyama 2001, 36). Unlike the concept of code used in semiotics to connect the sign to the meaning, resources enable us to explore and make sense of the different ways signs can be interpreted and assigned different meanings. Semiotic resources (such as the point of view of an image or the depth of focus in photography) are at the same time determined by the specific context in which they were created, and by the cognitive resources used to interpret images and their meanings. For this reason, the attention to semiotic resources implies attention to the ways the various 'rules' of interpretation came into being in a given cultural context, and the possibility of change in them.

Before moving to present visual social semiotics, a couple of considerations regarding semiotics resources are very important so as to use them appropriately as methodological tools of visual analysis. First, semiotics resources do not create meaning *per se*, but 'meaning potential': they make "possible to describe the kinds of symbolic relations between image producers/viewers and the people, place and things in the images" (Jewitt and Oyama 2001, 135). These meaning potentials are activated by the producers and the viewers of the images and do not, of course, convey a fixed meaning. However, they refer to a limited spectrum of meanings. Furthermore, it is important to keep in mind that the symbolic relations are indeed symbolic and very different from 'real' relations in the sense that their representation can purposely subvert real relations.

3.3   Visual Social Semiotics

Visual social semiotics is based on Michael Halliday's conceptualization of the three metafunctions of semiotic work: ideational, inter-personal and textual (Jewitt and Oyama 2001). The first has to do with the creation of representation, the second with the relation between the producer and the receiver of the text, and the last one to how these two functions work within their specific communication genre. Kress and Van Leeuwen (1996) have adapted Halliday's framework to the study of images and classified the three tasks of visual semiotics as representational, interactive and compositional. It is worth at this point presenting these three levels of analysis in detail because they will constitute the backbone of my analytical grid.

3.4   The Representational Meaning

The representational meta-function looks at the participants of the image, i.e. the people, object and places represented and, most importantly, at visual syntactic patterns that put the participant of the images in relation to each other. The structure dimension is important because it creates "meaningful propositions by means of visual syntax" (Kress and Van Leeuwen 1996, 47). The authors identify two kinds of representation: the narrative and the compositional. It is very important to notice that the choice among the two patterns is significant. For the choice to depict something in a narrative or conceptual way offers a "key to understanding the discourses which mediate their representation" (Van Leeuwen and Jewitt 2001, 141). In fact, visual structures do not simply mirror the structures of 'reality'. On the contrary, they create images of reality that are linked with the interests of the social institutions in which the images are created, disseminated, and used. "They are ideological. Visual structures are never merely formal: they have a deeply important semantic dimension" (Kress and Van Leeuwen 1996, 47).

3.4.1   Narrative Structure

The narrative structure refers to the way the different elements of the image are in relation one to another. The elements depicted are the *represented participants* – regardless of their humanity or non-humanity – and are distinguished from the *interactive participants*, namely producer and the viewer of the images. The relation among the represented participants can be of three types: transactional (characterized by the presence of a vector); locative (the contraposition between foreground and background given by the overlapping of shapes, the color saturation or the depth of focus), and instrumental (represented through the gesture of holding something). The main feature of narrative representation is the *vector*, a line that connects the various participants of the image. It can be represented by the position of a body, a hand pointing toward something, objects connecting represented participants (such as a weapon, camera, or toy) or eyelines.

Because narrative structures describe an action in its unfolding, the function of the vector of guiding the viewer through the narrative pattern is crucial and distinguishes narrative representation from conceptual ones that depict participants in their abstract meaning, in their essence. In photography, there are two kinds of represented processes: action or reaction. In action processes, the participants can be *actors* (from whom the vector, the action, generates), or *goals* (to whom the vector, the action, is directed). Whereas actors are always present in narrative pattern, goals can be absent. According to the presence of absence of a goal, we will talk, respectively of transactive or non-transactive action. When the vector is represented by eyeline, the process is of reaction. In this case the represented participants constitute the *reacters* and the object of their gaze "the *phenomenon*".

Another important aspect relating to the narrative structure is the different way through which participants can be put in relationship to each other in the image. Visual social semiotics individuates three types: conjoint (when participants are put in connection by a vector); compounded (when they are combined together but they have distinctive identities), and fused (when participants are fused together and their separate identities disappear). As Kress and van Leeuwen have pointed out: "each successive step further obscures the act of predication, the explicit act of bringing the two participants together, until the structure is no longer 'analytical', no longer analysed or analysable. We make the point at some length because of the (ideological) significance of this semiotic resource in configuring the represented world" (Kress and Van Leeuwen 1996, 53).

The analysis of the narrative structure includes the description of the settings, appearance of the represented participants, the props and the symbols present in the image.

3.4.2 Conceptual Structure

The conceptual structure represents the participants according to their general characteristic: "in terms of their more generalized and more or less stable and timeless essence, in terms of class, or structure or meaning" (Kress and Van Leeuwen 1996, 57). The authors identify three main kinds of conceptual representation: the classification, the analytical and the symbolic processes. The classification process refers to the representation of participants in a particular form or relationship to each other: that of taxonomy (which can be overt or covert according to the degree of explicitness of the overarching category), flowchart or network. The analytical process represents the relationship between the various parts and their whole structure: the parts are called *possessive attributes* and the whole *the carrier*. The analytical process is defined by the absence of vector, classification or symbolic process, and has a wide range of different structures such as temporal, topological or topographical, unstructured, exhaustive and inclusive. Finally, there are symbolic processes: the structures that represent the meaning of the participants.

These structures can be attributive (when the meaning of one participant, the 'carrier' is established through the meaning of the *symbolic attitudes*) or suggestive (when the 'carrier' represents the meaning in itself). Symbolic attributes are identified through their significant saliency, their position out-of-place, participants' gestures pointing at them, or their conventional social value. In suggestive symbolic structures, instead, the participant represents the meaning and differs from the analytical representation because of the de-emphazization of details and the use of modalities (see further on in this chapter) that maximize its generic quality and its timeless essence. Jewitt and Oyama (2001) have pointed out how this part of Kress and van Leeuwen's analysis draws from iconography and, as we will see, how iconography can complement visual social semiotics and be particularly helpful in identifying symbolic attributes and other visual motifs.

3.5   The Interactive Meaning

The interactive meaning is interested in grasping the relationship between the producer of the image and the viewer. Although their interaction can be direct and immediate (such as in the case of people taking pictures of each other as souvenirs), Kress and van Leeuwen note how the context of production and the context of reception are often disjoint. Disjunction aside, however, the producer and the viewer still share the image and "a knowledge of the communicative resources that allow its articulation and understanding, a knowledge of the way social interactions and social relations can be encoded in images" (Kress and Van Leeuwen 1996, 115). In visual communication, not only are social relations but also the relations between the producer and the viewer, represented, instead of enacted. This representation is created through different type of *resources*.

3.5.1   *Contact*

Some images establish a clear contact with the viewer. This is done through a vector (eyeline, or gesture) connecting the represented participants to the viewer. These kinds of images perform two key tasks: they both directly address the viewer and also constitute an "image act" (Kress and Van Leeuwen 1996, 117). Kress and van Leeuwen base the notion of "image act" on Halliday's concepts of "speech functions" that identifies four core speech acts and two reactions (expected and discretionary) for each: offer of information (social response: agreement or contradiction); offer of goods and services (social response: acceptance or rejection); demand of information (social response: answer or not answer), and demand of goods and services (social response: respond to the quest or not respond). When the gaze of the represented participant looks directly at the viewer, the producer is using the image to ask something of the viewer: an action, establishing a relationship, or creating an emotional bond. What kind of reaction the images are invoking depends on the details of the kind of look (perhaps probing, friendly or submissive) or gesture (perhaps inviting, defensive, or vexing).

On the contrary, when there is no eye contact, the images put the viewer in a *voyeuristic* position as unseen spectator. Following Halliday's classification, these images that do not address directly the viewer, are called *"offer images"* in contraposition to the images discussed above that belong to the *"demand"* category. In this case, the represented participants are offered to the viewer as "items of information, objects of contemplation, impersonally, as though they were specimens in a display case" (Kress and Van Leeuwen 1996, 119). As the authors make clear, these core types of images and a variety of sub-types and variation are possible along the contact resource spectrum. The function performed by the contact resource is, therefore, extremely important inasmuch as it indicates a specific kind of relationship between the viewer and the represented participant, suggesting with whom 'we', the viewer, should relate and who 'we' should just observe, and consequently who is the 'other'.

3.5.2 Distance

Distance is another way through which visual material depicts the relation between the viewer and the represented participants. Similarly to contact, distance is a term that refers to a continuum of the size of frame that can go from what is technically called a close-up to a very long shot. Drawing on the work of Edward Hall, Kress and van Leeuwen point out how, at the visual level, social distance is represented through the size of frame. A close shot corresponds to a close (or even intimate) social relation, whereas a very long shot corresponds to social distance. Visually this is represented along a continuum that goes from the depiction of only the head of a person to the portrayal of the full body (or bodies), including some headroom. In other words, the shorter the distance the stronger the connection, the social intimacy, with the represented participants and vice versa. In this sense, the authors' quotation of a painter, Grosser, is significant. The passage describes how the viewer will be forced to observe the 'soul' of the person portrayed at a distance of less than 90 cm while "at a distance of more than 13 feet (4m), people are seen 'as having little connection with ourselves', and hence 'the painter can look at his model as if he were a tree in a landscape or an apple in a still life'" (Kress and Van Leeuwen 1996, 125).

As in the case of contact, the distance is a powerful dimension of the interactive meaning. It creates, through a certain kind of representation, an imaginary relationship between the viewer and the represented participant, contributing to defining the people with whom we have a close or a far social distance and who are thus strangers to us. Although, as underlined above, representation is always about imagined relationships, not enacted or real ones, it is still extremely important to acknowledge its potential in creating a, more or less strong, social connection with the represented participants. As Kress and van Leeuwen have noted, as well as social distance, other important meanings (such as respect, objectivity, or authority) can be suggested by the conventional use of distance patterns: medium-close up with captions for the representation of experts speaking about an issue, close up for people telling their stories and diagrams for objective information.

3.5.3   Perspective

Perspective is another important dimension of the interactive meaning highlighted by Kress and van Leeuwen. This technique was firstly introduced in pictorial art during the Renaissance and used to represent depth and space on a two-dimensional surface. It provided the illusion of a stronger connection between reality and its representation and, at the same time, naturalized a point of view that was, on the contrary, socially determined.[1] Connected with the perspective and the concept of vanishing points (the points where the parallel lines seem to converge in a perspective image) is the notion of point of view, an important semiotic resource. The point of view indicates the position of the image producer toward the represented participants and the relationship among them thereby represented. It may have different angles, each of which represents power, involvement or detachment.

As with many of the other semiotic resources analysed so far, the angle of the image should be understood as a continuum of the whole range of possible points of view. Schematically, at the horizontal level, the image can have a frontal or an oblique angle. At the vertical level, a high angle represents a relationship of power of the viewer toward the represented participant while a low angle signifies the opposite and an eye-level angle a relationship of equality. Obviously, a wider range of nuanced meanings can be produced through all intermediate points of view within these visual perspectives extremes.

3.6   The Compositional Meaning

The compositional meaning refers to the way the representational and interactive meanings relate to each other and "the way they are integrated into a meaningful whole" (Kress and Van Leeuwen 1996, 176). Compositional meaning acquires, if possible, even more value in multimodal texts (texts that combine different semiotic modes such as written text and images), that comprise most of the data anlysed for this book. Kress and van Leeuwen consider three key elements of composition in

---

[1] For more on this point, see Kress and van Leeuwen 1996, 129–133.

their 1996 book and treat modality in a separate chapter. However, later on, Jewitt and Oyama (2001) include modality in this layer of analysis. Since all images considered in this study share the same modality (they are all photographic images), this aspect will require more attention. Although briefly introduced here, it will be the subject of a separate section (see 3.8 below).

3.6.1 Position and Information Value

The first element of composition is the position and the different information values Contained within the elements through their position in relation to the other elements. This dimension has to do with the represented participants' respective positions within the image, the respective position of two or more images, or the position that an image has with respect to text in a page (can be a newspaper page, or in the case of this study, a website page). The information value of different images is encoded into their left-right, top-bottom or center-margin positions and the three compositions can be found combined together. When represented participants or pictures are composed through a horizontal axis, the one positioned on the left will refer to what is 'given' – something unproblematic, agreed upon, self-evident – while the element of the right will refer to what is 'new', what is or should be at the center of attention, what is not known or agreed yet. Kress and van Leeuwen point out that although the statement made by this specific composition may be contested, or even denied, by the viewer, its ideological value lies in presenting the information in a particular way, conferring it a given new meaning. The second coding orientation confers different meaning according to the position at the top or the bottom of the composition. It usually presents less visual connection, if not even contrast, among the two elements. The image on the top represent the "ideal" – the promised situation, what might be – while the element on the bottom refers to the "real" – the situation how it is, empirics, sometimes even "directions for action" (Kress and Van Leeuwen 1996, 186). Finally, regarding the center-margin composition, putting an element at the center emphasises its core role and its predominance toward the element positioned around it. Of course, as the authors make very clear, those "coding orientations" are culturally determined and vary according to the diverse directionalities in different cultures. Thus, for example, in languages written right to left like Arabic the direction of images reading will follow the right to left orientation.

3.6.2 Salience and Framing

Another important element of the composition is the *salience,* that is the relative importance of the elements of the image. The more salient elements would be those that, by means of technical expedients, draw the attention of the viewer. Salience, as Kress and van Leeuwen explain, "is not objectively measurable, but results from complex interaction, a complex trading-off relationship between a number of factors: size, sharpness of focus, tonal contrast (e.g. high contract black and white images), colour contrasts (for instance, the contrast between strongly saturated and 'soft' colours, or the contrast between red and blue), placement in the visual field (elements not only become 'heavier' as they are moved towards the top, but also appear 'heavier' the further they are moved towards the left, due to an asymmetry in the visual field), perspective (foreground objects are more salient than background objects, and elements that overlap other elements are more salient than the elements they overlap), and also quite specific cultural factors, such as the appearance of a human figure or a potent cultural symbol" (Kress and Van Leeuwen 1996, 202).

The third element of the composition – framing – has to do with the degree to which the represented participants are connected, disjoined or separated to each other. A specific framing may be connectedness, discontinuity, or anything in the middle and is obtained through the use of colors, contrasts, white spaces, and vectors.

3.6.3   Modality

The last dimension of the compositional meaning is modality. This is defined and measured as the credibility or true value of the image. It does not imply the actual correspondence between representation and reality. Rather, it shows whether a visual element is represented as if it was true or not. The different levels of modality are obtained through so-called modality markers, or visual clues, that indicate how much we should trust the image. An extremely important point raised by social semiotics is the social construction of such modality markers. In other words, scholars have underlined how these visual clues "have arisen out of the interest of social groups who interact within the structures of power that define social life, and also interact across the systems produced by various groups within a society" (Kress and Van Leeuwen 1996, 155). The fact that what a social group considers real is culturally determined does not preclude the idea of realisms *per se* that will be, in turn, culturally determined. In this sense, the concept of realism has nothing to do with a factual correspondence between what is represented and the world. Rather, it is connected with the technological aspects of images production and hegemonic visual conventions. In our society, for example, the authors point out how photorealism is the "dominant standard" (Kress and Van Leeuwen 1996, 158) of realism. In photography, aspects such as color saturation, depth of field, amount of details contribute to the low or high modality of an image.

3.7 Iconography

Originally elaborated in the sixteenth century for the study of art, iconography was later developed and systematized in a three-level methodology for visual analysis by Erwin Panofsky (Müller 2011). The identification of visual motifs and interpretation of the meaning of visual products take place thought a three-step process: pre-iconographical description (or representational meaning according to the terminology used by van Leeuwen (2001); iconographical analysis (or iconographical symbolism) and iconological interpretation (or iconological symbolism). After a 'neutral' description of the represented elements, the second step is meant to identify typologies of images that share the same features. This categorization of images allows the researcher to recognize variances and resemblances that will – in the final step – be interpreted according to the wider social context. For the purpose of this study, the first and the second steps of analysis are particularly relevant in identifying visual motifs in the humanitarian discourse on the Syrian emergency and related migration crisis.

The first level, as with the denotation of visual semiotics and the representational meaning of visual social semiotics, refers to the description of the element of the image. Following Hermeren, van Leeuwen lists five ways to identify what is depicted: title or caption of the image; personal experience; background research; intertextuality, and verbal description. At the second level of analysis the represented participants – to continue with the terminology used by Kress and van Leeuwen (1996) – do not only denote the depicted individual/object, "but also the ideas or concepts attached to it" (Van Leeuwen 2001, 100). The attribute of iconicity refers to the resemblance of the image with the object that the image represents. In order to fully grasp the iconographical meaning of images it can be useful to keep in mind the distinction made by C.S. Peirce, one of the founders of semiotics, between icon, index and symbol (Peirce 1991). The first term refers to the similarity between the iconic sign and the object represented. Index is a sign clearly identifying this signified object. Symbols are images that conventionally (and therefore culturally specifically) establish a relationship between the representation and the object.

Although Panofsky initially elaborated the iconographical method in relation to art history, he recognized that the same pictorial conventions that connect concepts to artistic themes work in contemporary art. In the iconographical symbolism "there arose, identifiable by standardised appearance behaviour and attributes, the well-remembered types of the Vamp and the Straight Girl (perhaps the most convincing modern equivalents of the Medieval personifications of the Vices and Virtues), the Family Man and the Villain, the latter marked by a black moustache and a walking stick" (Panofsky quoted in Van Leeuwen 2001, 101).

Among the diverse images produced by humanitarian NGOs, iconography will allow the identification of a certain set of visual typification. The term, used by Kurasawa in an article on the iconography of humanitarian visuality, refers to a semiotic structure of images consisting of a relatively limited "system of formal relations between situational and compositional symbols serving to establish the roles of various actors (victims, perpetrators, aid workers, etc.) who are part of the visual composition of a scene of emergency or mass suffering" (Kurasawa 2015, 8). According to the author, the range of representations that are legitimate in a particular cultural, historical and socio-political context is limited, and its reiteration produces an "iconographic repertoire" of humanitarian images. The importance of visual conventions and repertoire lies in their being representative of a culturally and socio-historically situated system of thought, a way of representing the world that is shared by the practitioners who produce the visual material and works "as tacit referential or indexical social knowledge" (Kurasawa 2015, 20). The repertories play a pivotal role in the construction of the public discourse, setting the boundaries of how the people, situation and relations represented can be thought of and interpreted. Finally, the iconographical approach is even more interesting if we connect it with the argument of Heck and Schlag on the performativity of image in constituting an iconic act and how this understanding "directs our attention to the securitizing power of visual (re)presentations" (Heck and Schlag 2013, 896).

In his contribution on the political iconographic approach in the (Margolis and Pauwels 2011) (Eric Margolis and Luc Pauwels, eds., 2011), M.G. Müller suggests using previous literature on the research topic to identify typologies of visual motifs. For its specific attention to this aspect, iconography will be extremely useful to start identifying recurrent photographic patterns and attempt a first classification accordingly. Because of the massive diffusion of humanitarian images and their "iconic power" (Alexander et al. 2012; Kurasawa 2015) in the contemporary visual landscape, the literature is quite rich and provides numerous studies focusing on, inter alia, the iconography of suffering (Boltanski 1999; Chouliaraki 2013; Fehrenbach and Rodogno 2015); passivity (Nissinen 2015); personification, massification, rescue and care (Kurasawa 2015); piety (Shapiro 1988); emergency (Musarò 2017); humanitarian crisis (Campbell 2007); victimization (Friese 2017), or its opposite: resilient victim, and positive imaginary (Nissinen 2015), to cite but a few. In field humanitarian visual communication, iconography is also extremely useful to identify and describe the different iconological styles used by relief organizations. In the contemporary traditional and social media landscape, NGOs do compete at the visual level in order to distinguish themselves, their brand and their way of representing humanitarian issues (Kurasawa 2015). The identification of different iconographical approaches will be therefore extremely helpful in identifying the different organizational humanitarian narratives.

3.8 Photography, Power and 'Claims of Truth'

In discussing the visual approach of this study, it is also important to briefly discuss the relevance of the specific features of different visual genres and particularly the epistemic-political claims of photography, the visual genre object of the analysis. Probably nobody has expressed the importance of genre of communication more effectively than the sociologist Marshall McLuhan when he affirmed that "the medium is the message" (McLuhan and Fiore 1967). In the world of visual art, most people consider photography as a very specific medium, often opposed to other popular visual genre such as paintings or movies. Victor Bürgin (1982) has noted how the public usually receives pictorial art and films as objects that need to be experienced in a critical way, whereas photography presents itself as part of the environment. Similarly, Susan Sontag (1973) has shown how photography is commonly perceived as a transparent method showing a piece of reality, while writing and paintings are instead associated with interpretation.

The importance of the visual genre in creating the message is even more striking if we think in terms of what Hansen (2011) calls "epistemic-political constitution". Two aspects of this concept are particularly relevant for this study. The first has to do with the claim of what different genres do regarding their relationship with reality. In this sense, documentaries and photography are the two visual genres that derive their authority from their epistemic statement of truthfulness. The second level has to do with the degree to which different visual genre are expected to offer explicit political claims. It is not about an ontological political nature, but, rather, the expectation of the audience. In this respect, Hansen cites photojournalism and cartoons as the clearly political kind. However, the political claims of photography are probably more ambiguous.

It is important to discuss the complicated relation that this particular visual genre has with power and ideology. One of the most important aspects of photography is its relationship with reality. If all other forms of visual representation (such as pictorial art, sculpture or movies) represent, each in their respective peculiar way, an interpretation of the real world, photography is often thought of as the most objective way to catch reality (Sontag 1973). She clearly shows how photographic images have come to represent a miniature or reality, the testimony of a hidden truth, an instrument of knowledge, or a proof of reality (as proved by the fact that they have to be attached to some documents to make them valid). Similarly, Roland Barthes in *Camera Lucida* affirms that photography's "power of authentication exceeds the power of representation" (Barthes 1981, 89). According to Claude Levi-Strauss, this is particularly true for news photographs that "function as indexical illustration of the stories that accompany'" (Levi-Strauss quoted in Campbell 2007, 379).

Despite the ineluctability of aesthetics in any representation, pictures are commonly perceived as able to achieve an objective correspondence between the image and the referent object (Campbell 2007). It is exactly this assumption of photographs as 'unmediated simulacrum' that confers them so much authority in the field of "knowledge and truth" (Shapiro 1988, 124). Similarly, Annette Kuhn has pointed out how photographs imply authenticity and truth especially when what is represented through the lenses seems to be a credible surrogate of what we usually see. She notes that "the truth/authenticity potential of photography is tied in with the idea that seeing is believing. Photography draws on an ideology of the visible as evidence" (Kuhn 1885, 27).

This assumption of transparency confers photography authority and power. Since its inception, there have been two main, and yet opposite, perspectives regarding the relationship between photography and power. On one hand, photography has been praised for its ability to unveil and clarify. On the other hand – and this is the perspective that this study uses – it has been criticized for "its tendency to reproduce and reinforce the already-in-place ideological discourse vindicating entrenched systems of power and authority" (Shapiro 1988, 126). As Sontag (1973) contends in her seminal *On Photography*, to take a picture of something it is not only about appropriating what is represented, but also locating the image producer in a certain position toward the subject/object photographed, a position of knowledge and therefore power. Indeed, as she points out, photography has an inherently patronizing attitude toward reality resulting from its ambition to grasp the outside world and capture it through its lenses. The power relationship between the photographer and the subject can be looked at through different lenses: based on socio-economic class differences (Sontag 1973); neo-colonial approaches (Campbell 2007), or a gender perspective (Perna 2013). What is important here is that all these accounts share an acknowledgement that the photographic gaze presents itself as an objective eye, with "as if its perspective is universal" (*ibid.*, 42).

For the same reasons, photography is also strictly linked with ideology. As Bürgin has argued: "The structure of representation – point of view and frame – is intimately implicated in the reproduction of ideology (the "frame of mind" of our "point-of-view"). More than any other textual system, the photograph presents itself as *an offer you can't refuse*" (Bürgin 1982, 146). According to Shapiro, it is exactly the assumed truthfulness of photos that makes photography as a genre quintessentially ideological, obscuring the fact that "the *real* is forged over a period of time by the social, administrative, political, and other processes through which various interpretative practices become canonical, customary, and so thoroughly entangled with the very act of viewing they cease to be recognized as practices'" (Shapiro 1988, 185). Against this background, critical security studies have explored the difficulties linked to the seemingly opposite potentiality of visuality to repress and emancipate (Andersen et al. 2014). Scholars have shown how different visual genres have been associated either with the reproduction of domination and repression – such as in the case of mainstream film – or as constituting some form of resistance, in the case for example of artistic photography. They warned that at the theoretical level visual genre cannot be attributed *a priori* to emancipation, repression or critique capacities. Agreeing with the polysemic nature of images already underlined by Barthes (1981), they concluded, questions around their liberation/oppression potentialities can be only be answered at the empirical level. The advantage of this approach is to remain open-minded and admit that photography could sometimes contribute to the reproduction of hegemonic discourses, while at other times problematizing accepted analysis (Shapiro 1988).

The point is indeed to unveil how photographic enactment can reproduce power relations (Campbell 2007). Following Peirce's conceptualization of *icon, index* and *symbols* mentioned above, Campbell (2007) therefore suggests abandoning the understanding of documentary photography as *icons* and *indexes*, so as to fully acknowledge them as *symbols*. Or following Jean Baudrillard (1988), we could consider them as *simulacra,* that is to say the image's simulations of reality. Rather than a flattened and miniaturized version of reality, photographic images tell us something about the images producer, who unveil him/herself "through the camera's cropping of reality" (Sontag 1973, 95).

Drawing on this literature, the visual analysis of images examined in this book is inspired by a conceptualization of photography as a visual code, what Sontag defines as a "grammar and, even more importantly, an ethics of seeing" (Sontag 1973, 1). Photography is thus intended as "one signifying system among others in society which produces the ideological subject in the same movement in which they *communicate* their ostensible *content*" (Bürgin 1982, 153). To conclude, the point is therefore not to unveil the unfaithfulness of reality's representation, but, rather, to focus on the ways in which people and situations are enacted through their photographic depictions.

3.9 Polysemy and the Possibility of Different Readings

Before concluding the discussion of the visual approach, it is important to address a crucial feature of visual images, their polysemic value. Barthes has reminded us that "all images are polysemous; they imply, underlying their signifiers, a 'floating chain' of signifieds, the reader able to choose some and ignore others" (Barthes 1977a, 38–39). There is little doubt that when looking at the same image two people could be struck by different aspects of the representation. In a Barthian example of the advertisement in France of pasta *Panzani*, the level of denotation is clear and its reading as a quality food item in a shopping bag is quite straightforward to its audience. But at the level of connotation, the same image lends itself to multiple meanings: sign of freshness (just returned from the market); domestic preparation of food; *Italianicity*, the "idea of a total culinary service" since everything needed for a meal seems to be there, and the evocation of pictorial still life (Barthes 1977b, 270–71). Polysemy should be therefore taken seriously into account in any visual analysis.

However, the various readings are somehow circumscribed as "in every society various techniques are developed intended to *fix* the floating chain of signified in such a way as to counter the terror of uncertain signs" (Barthes 1977a, 39). In the interpretation of images, therefore, the meanings circulating among a situated cultural milieu assume particular importance in limiting the various reading possibilities. In this sense, as Mitchell has maintained, "whatever the pictorial turn is, then, it should be clear that it is not a return to naïve mimesis, copy or correspondence theories of representation, or a renewed metaphysics of pictorial "presence"; it is rather a postlinguistic, postsemiotic rediscovery of the picture as a complex interplay between visuality, apparatus, institutions, discourse, bodies, and figurality" (Mitchell 1995, 4–5).

In choosing to interpret images through the specific methodology of visual social semiotics, I am also aware that this is just one of the ways through which an image can be read and its meaning unpacked. Far from implying that this is the *only* or the *right* way to analyse a visual artefact, I am suggesting that this particular perspective is worthy of exploration for two main reasons. First, beyond its attention to the wider cultural and social context, visual social semiotics' interpretation of an image is based on the complex agglomeration of the multiple semiotic resources at play and the interplay of the different layers of meaning (i.e., representational, interactive, compositional). In each picture, this infinite possibility of combination works to reinforce or, on the contrary, weaken, a particular reading. Consequently, an image is analysed in its entirety and, in each case, the various layers of meaning and semiotic resources considered together can help point towards one specific reading. A reading, it goes without saying, that is situated in a geographically and historically specific cultural milieu. This is precisely the second and most important aspect. The analysis is based on a specific situatedness that is linked with my positionality as part of the Western contemporary audience, which is exactly the one to which the images which are the objects of this study are directed to. Since any reading is situated in a particular cultural milieu, as Barthes has noticed, cultural and social expectations are brought to the image. Being part of the same cultural milieu, which is the primary audience of the image, is therefore crucial to unpacking the various meanings that are possible in that specific culturally, geographically and historically situated public. Moreover, the visual analysis has been complemented by a multi-sited fieldwork, direct engagement and investigation of the images' producers – the transnational humanitarian NGOs. In so doing images have been analysed keeping in mind a much wider set of contextual information. Even acknowledging that that interpretation is just one among the multiple possible, it is important as it unveils one of the meanings that a picture assumes in a certain cultural milieu in a given moment in time and space. Highlighting that specific interpretation does not mean providing a unilateral and deterministic meaning assignment, but, rather, to unpack a certain reading and stimulate discussion of the relevance that that reading has on a specific public.

References

Alexander, J., Bartmanski, D., & Giesen, B. (2012). *Iconic power: Materiality and meaning in social life*. New York: Springer.

Andersen, R. S., & Möller, F. (2013). Engaging the limits of visibility: Photography, security and surveillance. *Security Dialogue, 44*(3), 203–221.

Andersen, R. S., Vuori, J., & Mutlu, C. E. (2014). Visuality. In C. Aradau, J. Huysmans, A. Neal, & N. Voelkner (Eds.), *Critical security methods: New frameworks for analysis*. London: Routledge.

Andersson, R. (2012). A game of risk: Boat migration and the business of bordering Europe (Respond to this article at http://www.therai.org.uk/at/debate). *Anthropology Today, 28*(6), 7–11.

Barthes, R. (1972). *Mythologies. 1957* (Annette Lavers, Trans., pp. 302–306). New York: Hill and Wang.

Barthes, R. (1977a). *Image music text*. London: Harper Collins Publishers, Fontana Press.

Barthes, R. (1977b). *Rhetoric of the image*. New York.

Barthes, R. (1981). *Camera Lucida: Reflections on photography*. New York: Macmillan.

Baudrillard, J. (1988). *Baudrillard*, Selected Writings, Ed. Mark Poster.

Bleiker, R. (2001). The aesthetic turn in international political theory. *Millennium, 30*(3), 509–533.

Bleiker, R., & Kay, A. (2007). Representing HIV/AIDS in Africa: Pluralist photography and local empowerment. *International Studies Quarterly, 51*(1), 139–163.

Blommaert, J., & Bulcaen, C. (2000). Critical discourse analysis. *Annual Review of Anthropology, 29*(1), 447–466.

Boltanski, L. (1999). *Distant suffering: Morality, media and politics*. Cambridge, UK: Cambridge University Press.

Bürgin, V. (1982). Looking at photographs. In *Thinking photography* (pp. 142–153). New York.

Campbell, D. (2003). Cultural governance and pictorial resistance: Reflections on the imaging of war. *Review of International Studies, 29*(S1), 57–73.

Campbell, D. (2007). Geopolitics and visuality: Sighting the darfur conflict. *Political Geography, 26*(4), 357–382.

Chouliaraki, L. (2013). *The ironic spectator: Solidarity in the age of post-humanitarianism*. Hoboken: Wiley.

Derrida, J. (1967). La Structure, Le Signe et Le Jeu Dans Le Discours Des Sciences Humaines». In *L'Ecriture et La Différence, Paris, Seuil* (pp. 409–428).

Fairclough, N. (1992). Discourse and text: Linguistic and intertextual analysis within discourse analysis. *Discourse & Society, 3*(2), 193–217.

Fehrenbach, H., & Rodogno, D. (2015). "A horrific photo of a drowned Syrian child": Humanitarian photography and NGO media strategies in historical perspective. *International Review of the Red Cross, 97*(900), 1121–1155.

Foster, H. (1988). *Vision and visuality*. Seattle: Bay Press.

Foucault, M. (1971). Orders of discourse. *Social Science Information, 10*(2), 7–30.

Foucault, M. (1972). *Archaeology of knowledge*. London: Routledge.

Foucault, M. (1980). *Power/knowledge: Selected interviews and other writings, 1972–1977*. New York: Pantheon.

Friese, H. (2017). Representations of gendered mobility and the tragic border regime in the Mediterranean. *Journal of Balkan and Near Eastern Studies, 19*, 1–16.

Hansen, L. (2011). Theorizing the image for security studies: Visual securitization and the Muhammad cartoon crisis. *European Journal of International Relations, 17*(1), 51–74.

Haraway, D. (1991). *Simians, cyborgs, and women: The reinvention of nature*. New York: Routledge.

Heck, A., & Schlag, G. (2013). Securitizing images: The female body and the war in Afghanistan. *European Journal of International Relations, 19*(4), 891–913.

Jewitt, C., & Oyama, R. (2001). Visual meaning: A social semiotic approach. In *Handbook of visual analysis* (pp. 134–156). Los Angeles: Sage.

Kress, G. R., & Van Leeuwen, T. (1996). *Reading images: The grammar of visual design*. New York: Psychology Press.

Kuhn, A. (1885). *The power of the image: Essays on representation and sexuality*. London: Routledge.

Kurasawa, F. (2015). How does humanitarian visuality work? A conceptual toolkit for a sociology of iconic suffering. *Sociologica, 9*(1), 0–0.

Margolis, E., & Pauwels, L. (2011). *The Sage handbook of visual research methods*. Los Angeles: Sage.

McLuhan, M., & Fiore, Q. (1967). *The medium is the message* (Vol. 123, pp. 126–128). New York: Bantam Books.

Mitchell, W. J. T. (1995). *Picture theory: Essays on verbal and visual representation*. Chicago: University of Chicago Press.

Möller, F. (2007). Photographic interventions in Post-9/11 security policy. *Security Dialogue, 38*(2), 179–196.

Müller, M. G. (2011). Iconography and iconology as a visual method and approach. In *The SAGE handbook of visual research methods* (pp. 283–297). London: Sage.

Musarò, P. (2017). Mare Nostrum: The visual politics of a military-humanitarian operation in the Mediterranean Sea. *Media, Culture and Society, 39*(1), 11–28.

Nissinen, S. (2015). Dilemmas of ethical practice in the production of contemporary humanitarian photography. In Rodogno & Fehrenbach (Eds.), *Humanitarian Photography* (Vol. 301). Cambridge: Cambridge University Press.

Nöth, W. (2011). Visual Semiotics: Key features and an application to picture ads. In *The Sage handbook of visual research methods* (pp. 298–316). Los Angeles: Sage.

Peirce, C. S. (1931). *Collected papers of Charles Sanders Peirce*. 8 Volumes (Eds. Charles Hartshorne, Paul Weiss & Arthur W Burks, Vols 1–6).

Peirce, C. S. (1991). *Peirce on signs: Writings on semiotic*. Chapel Hill: UNC Press Books.

Perna, R. (2013). *Arte, Fotografia e Femminismo in Italia Negli Anni Settanta*. Postmedia books. https://scholar.googleusercontent.com/scholar.bib?q=info:ET_2l3bygxIJ:scholar.google.com/&output=citation&scisig=AAGBfm0AAAAAWVqvMfmifcI5YTwrbPtEeM1ocoAy6nev&scisf=4&ct=citation&cd=-1&hl=it

Robinson, P. (1999). The CNN effect: Can the news media drive foreign policy? *Review of International Studies, 25*(2), 301–309.

Rose, G. (2001). *Visual methodologies: An introduction to researching with visual materials*. London: Sage.

Semiotics Encyclopedia Online. (2018). *Social semiotics*. In E.J. Pratt Library. Victoria University. https://semioticon.com/seo/S/social_semiotics.html

Shapiro, M. J. (1988). *The politics of representation: Writing practices in biography, photography, and policy analysis*. Madison: University of Wisconsin Press.

Shapiro, M. (1999). *Cinematic political thought: Narrating race, nation, and gender*. New York: NYU Press.

Shapiro, M. J. (2007). The new violent cartography. *Security Dialogue, 38*(3), 291–313.

Slembrouck, S., Verschueren, J., Ostman, J. O., & Blommaert, J. (1995). Channel. *See Verschueren et Al* 1995, pp. 1–20.

Sontag, S. (1973). *On photography* [Online] (Electronic edition 2005). New York: Rosetta Books LLC.

Van Dijk, T. A. (2015). Critical discourse studies: A sociocognitive approach. In *Methods of critical discourse studies* (pp. 63–74).

Van Leeuwen, T. (2001). Semiotics and Iconography. In *Handbook of visual analysis* (pp. 92–118). Los Angeles: Sage.

Van Leeuwen, T., & Jewitt, C. (2001). *The handbook of visual analysis*. Los Angeles: Sage.

Wang, J. (2014). Criticising images: Critical discourse analysis of visual semiosis in picture news. *Critical Arts, 28*(2), 264–286.

Weber, C. (2006). *Imagining America at war: Morality, politics and film*. London: Taylor & Francis.

Weldes, J. (2006). *To seek out new worlds*. New York: Springer.

Wodak, R. (2001). What CDA us about-a summary of its history, important concepts and its developments. In R. Wodak & M. Meyer (Eds.), *Methods of critical discourse analysis*. London: Sage.

Wodak, R., & Meyer, M. (2009). *Methods for critical discourse analysis*. London: Sage.

Part II
Humanitarian Representation and Migration Governance

Chapter 4
Humanitarian NGOs and Global Governance: One, No One and One Hundred Thousand Humanitarian NGOs

4.1 Introduction

To understand the role that NGOs' representation of Syrian displacement plays in global governance before getting into the visual analysis it is important reflect upon the aspirations of emergency organizations. How do relief agencies intend, perceive and present their role to the public? Are they interested in participating or influencing global governance? Do they consider their role as promoters of universal values or technical agents performing a specific task? Answering these questions is important to unpack their distinctiveness and the different ways in which different NGOs conceive and perform their mission in the international arena. In this sense, it is extremely interesting to look at how relief organizations accommodate their humanitarian role and the humanitarian principles within contexts that are inescapably highly political (e.g., situations of violence, displacement, political contestation or belligerent occupation). Not only do NGOs work within a complex web of political interests, international relations and systems of power, but, for better or worse, their humanitarian and advocacy actions have practical political implications. The investigation of how different organizations negotiate their relationship with politics allows us to better understand where each positions itself within the heated debate around the interrelations of humanitarianism and politics discussed in the first chapter.

Barnett has brilliantly argued there have been various indications that NGOs: "might themselves be helping to reproduce and expand the very world order that they claim to resist (…) Certainly many humanitarian agencies do, worrying that their agendas are serving the interests of powerful states , that they are furthering a liberal world order that advantages some over others, and that they are part of governance structures that place them in positions of power over the very people they claim to want to emancipate" (Barnett 2009, 656).

© The Author(s) 2021
A. Massari, *Visual Securitization*, IMISCOE Research Series,
https://doi.org/10.1007/978-3-030-71143-6_4

Yet, at the same time, he points out, these organizations have also shown ethical agency and some degree of reflexivity about their relationship with systems of power, politics and global governance. NGOs have modified their policies, adapted practices, and tried to develop alternative ethics.

Within each organization, there are many different aspects indicative of the space in which they operate: financial (in)dependence from governments; purely humanitarian versus humanitarian-development mandate; needs-based versus rights-based approaches to humanitarian assistance or the objectives of their advocacy strategies. Identifying the different strategic objectives NGOs prioritize (whether psychosocial support, medical assistance, women empowerment, education, or giving a voice to the powerless, to name but a few), allows us to understand which kind of humanitarian role organizations want to play in relation to people they are assisting and societies in which they intervene.

This chapter introduces the four major transnational NGOs – Save the Children, Oxfam, CARE and MSF – outlining their respective histories, current institutional organization, geographical scope and the origin of their funding. It will also outline the role that each has designed for itself by looking at the extent of their mandate, their vision, how their staff reflect on the complicated relationship of their organizations' humanitarian work with politics and global governance, and the key objectives of their advocacy strategies. Finally, the chapter will present NGO-specific interventions in the emergency response to Syrian displacement, both in terms of humanitarian action and advocacy.

4.2   Save the Children

The global organization today known as Save the Children was founded in post-World War I London by Eglantyne Jebb. She was keen to draw public attention to the suffering of children in Austria-Hungary and Germany, defeated countries enduring an Allied blockade. She established the Save the Children Fund in 1919 and in 1924 drafted a document that was adopted by the League of Nations as the *Declaration of the Rights of the Child*. During World War II Save the Children, continued its focus on the protection of children. Inspired by Jebb, the first humanitarian to use sophisticated fund-raising techniques, local national branches were founded. In 1977, they came together to form an alliance. Today SCF is comprised of 30 national member organisations and a Save the Children International secretariat. Save the Children works in some 120 countries across the globe, primarily engaged in humanitarian relief, health, nutrition, water and sanitation, education, child protection and safeguarding, the rights of the child, advocacy and child poverty.

The website of SCF International and official documents do not explicitly mention humanitarian principles, but they refer to the organization's adherence to the Core Humanitarian Standards (CHS) (CHS 2020)– the essential elements of principled, accountable and high-quality humanitarian action which are based on humanitarian principles. Moreover, the organization adheres to Sphere standards for humanitarian action and promotes quality and accountability (Sphere 2020). Save the Children further endorses Code of Conduct for the Red Cross and Red Crescent Movement and NGOs in Disaster Relief (IFRC 1994), which make clear reference to the humanitarian principles.

At the time of the research, Save the Children International and the 30member organizations in 2015 reported a total (annual) gross income of 1.9 billion Euro (Save the Children 2016b).[1] Nearly two third (58%) came from institutions, including Governments, the rest from private donations (25%), corporations and foundations (13%), gifts in kind (2%) and other income sources (2%) (Save the Children 2016b). The 2016 Global Strategy (Save the Children 2016a) stated among its priorities a desire to "Drive stronger, more diversified funding" so as to raise the percentage of unrestricted funding – funds that are not bound to any specific program and that the agency can use at its own discretion – from 21% of global income in 2014 to 25% in 2019, along with increasing the donations from Middle Income Countries. According to Save the Children's Annual Report (Save the Children 2016b), the NGO's expenditures in 2016 amounted to 44% of the global income for humanitarian programmes and 56% for development programs.

From a quick look at these data, it emerges quite clearly how from the specific objective to alleviate suffering in a particular historical and geographical context, the organization has grown to have a global presence with global objectives encompassing a very wide range of interventions. With a physical presence in some 120 countries Save the Children is a transnational organization whose scope goes today well beyond the initial goal of feeding hungry children in countries who lost World War I.

4.2.1 Save the Children and Society

Although present in several emergency situations around the world, Save the Children's work is not "only" humanitarian: the organization in facts works both on development and humanitarian programmes. Its Global Strategy *Ambition for Children 2030* is explicit about it: "millions of children are being left behind and denied the opportunity to survive and fulfil their potential. This document outlines the change we want to inspire by 2030, as well as the strategic steps we will take over the next three years to get there. We will do whatever it takes to ensure by 2030 all children survive, learn and are protected, with a focus on the most deprived and marginalised children" (Save the Children 2016c, 3). While *survival* and *protection* mainly refer to the humanitarian objectives of saving lives and alleviating suffering, the objectives referring to children learning and fulfilling their potential have more

---

[1] The official figure is $2.1 billion. To make it comparable to the other organization's income in euro, the exchange rate of Inforeuro in December 2016 has been applied.

to do with development work. The idea is to intervene in an comprehensive manner and provide people with 'an integrated response to their needs: immediate humanitarian relief, long-term solutions and continued campaigning for an end to conflicts" (Save the Children 2018e).

Save the Children's holistic approach is reaffirmed in its vision of "a world in which every child attains the right to survival, protection, development and participation' and the organization's mission to 'inspire breakthroughs in the way the world treats children and to achieve immediate and lasting change in their lives" (Save the Children 2018f). Expressions such as *change the world, transform lives* and *building a better world* populate the document, and together with the organization's embracing of the rights-based approach (Save the Children 2005), clearly indicate intention to work for long-term goals, well beyond immediate humanitarian assistance. These expressions, however aspirational they may be, can help to understand the role that the NGO conceives for itself in global governance: changing the world and transforming lives. While SCF was established to alleviate the suffering of children today its objectives include a much wider set of actions aimed at improving child rights governance around the world.

Save the Children's goal to affect social and policy changes on children rights (in particular its advocacy for universal ratification of the *UN Convention on the Rights of the Child*) is presented as a progressive and innovative approach meant to overcome the limitations often associated with traditional, at times considered shortsighted if not unsustainable, humanitarianism. The interaction of humanitarianism with policy reform, and inevitably with its dynamics of power and politics, is considered, and presented, in positive terms, as an important opportunity for Save the Children to have long-term transformative results for on the governance of children rights.

At the same time, and quite oddly, the interrelation of humanitarian work with politics is also considered by the organization as something inherently negative. Even more, avoiding any kind of reference and potential interaction with the political sphere appears more functional for the NGOs' achievement of transformative objectives. In my interviews with Save the Children's staff several informants underlined that the organization did not have anything to do with politics. At the same time, however, all the people I spoke with affirmed that for the NGO performing a role in global governance was extremely important. This widespread opinion was clearly summarized in the words of a Save the Children advocacy and communications manager: "We avoid politics by sticking to the evidence. For example, 50 per cent of Syrian children do not have opportunity to education. If we do not use research, figures, they [national policy makers] will think that we are biased, and we are supportive of refugees. We will instead say that children without education are more prone to be attracted by armed groups, terrorist organizations, drug abuse, etc. We will try to make the policy makers understand that ensuring education for children is in the best interest of Lebanon. We will also say that all support for refugees benefits all of society (not only the refugees). We will not present positions but thorough facts that can influence the government position ...We take many factors into account: the donors, the will of the government, needs, and facts."[2]

Despite the irony of an extremely political statement made precisely to avert any implication with politics, it is very interesting to notice how from the NGO perspective politics can be avoided. Implicit in this affirmation, there is also the suggestion that there is a neutral objective reality that NGOs can appeal to in order to remain outside of the political debate. This is a very political a-political statement.[3] The words of the practitioners indicate that the organization relies on the technicality of its work as something objective and disconnected from the prevalent historical and political environment. Despite the strategic intention of tackling violations of children rights' through a holistic approach taking into consideration the root causes of a given situation, this statement also reveals the assumption that the NGO consider, and present, its role as technical and objective, and that this self-proclaimed a-political stance can be functional in obtaining politically impactful policy and societal transformations.

Within the strategic objective of achieving impact in changing children' lives, advocacy plays a crucial role. As the Save the Children international website states: "for Save the Children, advocacy is the work we do to influence the policies and actions of governments, international institutions and the private sector, in order to achieve positive changes in children's lives (…) In different situations, our advocacy can be focused on securing formal policy changes; driving implementation; or creating an enabling environment for change" (Save the Children 2018a).

Advocacy work is strictly connected with another Save the Children's sector of intervention that the organization itself defines as *Child Rights Governance*. Under this ambitious title, the NGO works to ensure that governments fulfil children's rights. In practice, this means influencing policy makers to keep children rights at the top of their agenda and promote policies and law changes to ensure that the rights of children are respected. In the organization's words: "In 2014 alone, our work contributed to governments spending more on children in 12 countries, and we helped 18 countries to change policies or laws to make child rights a reality for more children (…) We aim to raise the issue and status of children in all societies to ensure that they are treated as the citizens of today not just tomorrow and in doing so push children and their rights up the political agenda" (Save the Children 2018b).

The crucial role that advocacy plays in the organization's ambition to affect global governance emerges clearly from the fact that the overarching Save the Children's objective to improve child rights governance is primarily based on advocacy and campaigning. Many informants stressed that because of the importance of advocacy within the agency's global strategic objectives, and to ensure capacity to work at different level of governance, each national branch has an advocacy

---

[2] Interview with Save the Children Media Advocacy and Communications Manager, 8th March 2017, Beirut.

[3] I thank Anna Triandafyllidou for this insightful comment.

department at the country level to deal with national issues, and regional and international departments to work respectively at the regional and international level.

Save the Children's advocacy significantly depends on celebrity supporters (described by SCF UK as Artists and Ambassadors). Ilan Kapoor has shown how "celebrities humanitarianism" is intrinsically contaminated and ideological, rather than altruistic. It is problematic not only insofar as its promotes consumerism and corporate capitalism and reifies the north-south inequalities it was intended to address in the first place, but also because it contributes to the depoliticization of global inequalities (Kapoor 2012). The use of celebrities as "Northern saviours" (Van den Bulck 2009; Duvall 2009), most strikingly exemplified by iconic imagery of Princess Diana in the minefields of Angola or Angelina Jolie in the camps of Darfur, is not something new in humanitarian communication. Chouliaraki has noted how celebrity advocacy plays on the level of emotions, by theatrically exposing the celebrities' "authentic"' feelings in a way that we can connect with, thereby encouraging "a narcissistic disposition of voyeuristic altruism rather than one of commitment to the humanitarian cause" (Chouliaraki 2012, 17).

In common with many other major humanitarian actors, Save the Children has in recent years increasingly focused on partnership with private capital. Through what it terms its Global Corporate Partnerships, the organization works "toward positively addressing the unique and complex challenges facing children, their families and communities, while simultaneously creating value for partners' businesses and brands" (Save the Children 2016a, 8). Such collaboration brings together "marketing teams on cause-aligned campaigns and licensing programmes that leverage the power of our combined brands and consumer reach" to create "industry coalitions on mutual areas of interest, bringing collective support to large-scale projects"(Save the Children 2016a, 8). As touched on in the first chapter, inherent problems arise in such collaboration and marketization and the problematic interplay between humanitarian aspirations and business objectives. The implication of combining business goals – and their inherent interest in economic profit – with relief intervention raises questions about the extent to which humanitarian objectives need to be negotiated and accommodated to allow such collaboration. The collaboration with private capital is not only an organizational strategy that can benefit NGOs or UN agencies. What needs to be grasped are the implications that such interplays can have on the practical consequences that humanitarian interventions have in the context in which they occur.

4.2.2 Save the Children and Syrian Displacement

Save the Children, which has had programs in the Middle East for decades, has been strongly involved in the humanitarian response to the Syrian war. The organization has been operational both within Syria and in neighboring countries of refuge (principally Jordan, Lebanon, Iraq and Egypt), along the eastern Mediterranean route, popularly known as the Balkan route (i.e., Turkey, Greece, Serbia Croatia, Italy),[4] into preferred countries of asylum in northern Europe (e.g., the Netherlands, Germany and Switzerland), and in the Mediterranean.

Between 2015 and 2016, the bulk of Save the Children work in countries hosting Syrian refugees mainly focused on child protection and education. The organization established several child and youth friendly spaces – both in camps and in urban areas – where children were offered recreational activities and psychosocial support services to help cope with the traumatic experience of war and displacement. Save the Children supported host governments to offer formal education to eligible Syrian refugees. For those children who could not enter the formal education system (because of lack of documentation or financial resources) the organization provided non-formal and remedial education. Though not a major player, Save the Children also distributed food and basic relief items, provided cash assistance, organized activities seeking the economic empowerment of vulnerable hosting communities and implemented water and sanitation, and shelter rehabilitation projects.

Although at first sight quite technical activities, different interventions have important implications for the way in which NGOs participate in global governance and particularly, as in this case, how they contribute to the shaping of the governance of refugees. For example, Save the Children's emphasis on psycho-social support activities is part of a logic that Pupavac (2001b) has defined as '"therapeutic governance". Psycho-social interventions, culturally deeply rooted in the Anglo-American therapeutic ethos, are centered on social risk management strategies. The basic assumption of the psycho-social model is based on the vulnerability of the individual, and particularly the vulnerability of women and children, with quite limited attention to adult males and with scant regard for the specific social roles in war and displacement. Refugees, either seen as traumatized or alternatively in denial of their trauma and susceptible to be re-traumatized by memories, are inevitably considered dysfunctional. "As a consequence of the pathologisation of their condition, psycho-social intervention implicitly denies the capacity of populations for self-government" (Pupavac 2001b, 365) and may severely undermine local coping strategies. Moreover, the pathologisation of war-affected communities implies refugees' disqualification as political actors, while at the same time "validating the role of external actors", namely the humanitarian organizations (Pupavac 2001b, 367). This perspective is important not only because it highlights how the psycho-social model resonates with colonialist approaches characterized by the pathologization of the dependent subject, but also for its insight into the focus on children that Save the Children shares with many other organizations. Drawing on Burman's work, Pupavac (2001a) has pointed out how the emphasis on children has contributed to the reproduction of a stereotyped idea of an adult professional from the Global North helping and supporting an infantilized subject in the Global South.

---

[4]The eastern Mediterranean route was the most-used pathway for Syrians attempting to reach Europe between 2015 and 2016. For more information on what as became popularly known as the Balkan route, see Achilli 2016.

In 2016, in response the growing number of people losing their lives trying to cross the Mediterranean en route to Europe, Save the Children established its own maritime SAR operation, chartering a vessel which operated until SAR was abandoned following repeated clashes with the Libyan coastguards. Although as Cuttitta (2017) has observed, SAR operations have a substantial impact in both the depoliticization and repoliticization of the EU maritime border, Save the Children's political positioning on its SAR activities was one of political neutrality, the agency investing much effort in trying to keep politics separate from its life-saving activities.

In addition to its work in the field, the organization is also committed to advocacy and global campaigning. Syria is one of the sites of Save the Children's biggest campaign, *Every Last Child,* that focuses on the world's excluded children, with a campaign website dedicated to the children affected by the Syrian crisis (Save the Children 2018g). Different operational branches engaged in several social, communication and fundraising campaigns specifically focusing on Syria. These included: *child refugee crisis* (Save the Children 2018c), *the space migrant* (Save the Children 2018d), *most shocking second a day* (Save the Children 2014), *still the most shocking second a day* (Save the Children 2016d), and numerous hash tags such as #iostoconAylan, #NotJustARefugee, #savechildrensatsea, #SaveChildRefugees.

Save the Children also played a major role in the various fora gathering together international organizations working on the Syrian response and particularly in their advocacy working groups. The organization has been a signatory of several collective NGO advocacy documents such as *Failing Syria* (Various NGOs 2015) and *Stand and Deliver* (NGOs Platforms et al. 2017). In relation to the so-called European refugee 'crisis', SCF produced an advocacy report summarizing the key priority recommendations at EU, regional and national levels (Save the Children 2017). Politics is quite clearly called into question in the organization's advocacy strategy as those documents address the context of the war. However, policy recommendations are presented in a consciously apolitical and technical discourse

4.3 Oxfam

Oxfam International was established in 1995 by a group of NGOs. The name Oxfam emerged from an initiative in 1942, the largely Quaker-inspired creation of the Oxford Committee for Famine Relief, founded in 1942 advocating for food supplies to be sent to women and children suffering famine in war-torn Greece (Barnett and Weiss 2008) The organization rose to prominence during the war for Biafran independence at the end of the 1960s when it chose to disregard Nigeria's land and sea blockade of Biafra and organized an airlift. In striking contrast with the policy of the International Committee of the Red Cross (ICRC) to remain neutral and to respect the sovereignty of the Nigerian state, OXFAM took an explicit partisan position affirming that "the price of a united Nigeria is likely to be millions of lives" (De Waal cited in Chandler 2002, 30). Today Oxfam International is a confederation of 20 affiliated organizations around the world and maintains an international secretariat in Oxford. It works in 90 countries across five continents, focusing on fighting poverty and inequality through projects based on active citizenship, natural resources, women rights, access to services, life-saving activities and sustainable food (Oxfam International 2018d).

The organization defines itself as independent and impartial according to the following definitions adapted by the ICRC: "*Impartial*: for every person, according to their needs, without discrimination because of race, gender, religion, age or anything else; and *Independent*: directed without influence from any interest group or political group' (Oxfam International 2013, 2). Oxfam, like SCF and other major humanitarian players also is committed to Sphere standards, *The Code of Conduct for the Red Cross and Red Crescent Movement and NGOs in Disaster Relief,* and the *Core Humanitarian Standards*. It also, like other NGOs, has signed up to Do no Harm Principles, stating that "the humanitarian imperative does NOT mean that we (…) should provide assistance without considering whether its harm could outweigh its benefit. We must judge the likely immediate and long-term consequences of different actions before deciding how, and whether to provide aid' (Oxfam International 2013, 2).

On neutrality, Oxfam presents a particular perspective as it links the term to its advocacy role and the possibility to speak out in cases of extreme suffering or violation of people's rights. In the words of the organization: "impartial advocacy does not mean saying that every party to violence is always equally to blame. Nor is Oxfam neutral in the sense of avoiding anything that could be construed as a policy controversy. We take a stand on the causes of humanitarian need and propose policy changes to solve them – based on our experience, values, and international humanitarian law. Speaking out Oxfam routinely bears witness to extreme suffering and violations of people's rights under international humanitarian, refugee and human rights law. This is part of our responsibility to raise the voice of those affected, alert the world, and call on relevant authorities to take action. Usually we do that in public, but sometimes that may create unacceptable risks to the safety of our staff and others, or to our ability to provide assistance. Judging when and how to speak out is never easy. Campaigning can be vital to ensure people can reach the aid that they need. But we may also have to negotiate with parties to a conflict, in order that affected men, women and children can reach the assistance they need. And that may require us to limit our campaigning or to support other organisations rather than 'speak out' as Oxfam ourselves" (Oxfam International 2013, 3).

Oxfam's major source of income is from institutions (41.4%), followed by public f (39.6%), trading revenue (17.5%), and other incomes (1.1%) for a total of 1.05 billion Euro. Within the expenses relating to programme implementation, 52% is spent on development and humanitarian activities, and 6% on influencing activities (Oxfam 2017b).

Like Save the Children, Oxfam has considerably expanded from its original objective of addressing famine. The agency has sought to position itself as willing to oppose state policies. This self-perception and self-positioning of Oxfam is indicated by its emphasis on humanitarian principles and particularly on its vision of neutrality and the need to speak out. From a financial perspective, Oxfam is not as independent as it likes to imply it is. Unlike, for example, MSF, Oxfam is heavily dependent on UK and EU funding. This accounted for 43% of income in 2017.

4.3.1 Oxfam and Society

Oxfam's Strategic Plan (Oxfam 2013) lists the organization's key priorities: "change our world", "tackle poverty", and foster "lasting resilience to crisis and poverty". Central to Oxfam's approach is the concept of "transformational change", the long-term goal to address the root causes of poverty through structural practices policy changes. Oxfam's vision is of "a just world without poverty (…) a world where people are valued and treated equally, enjoy their rights as full citizens, and can influence decisions affecting their lives" (Oxfam 2017a). In order to achieve this, Oxfam seeks to "create lasting solutions to the injustice of poverty. We are part of a global movement for change, empowering people to create a future that is secure, just, and free from poverty" (Oxfam 2017a).

It is quite clear from the organization's own statements how the agency sees its role as key in global governance. Its position is clearly summarized in the section outlining the organization's beliefs: "We face unprecedented changes and challenges this century, including climate change, famines and food price crises, increasing humanitarian crises, energy limitations, proliferation of weapons, urbanization, and natural resources shortages. To meet these challenges, we need global cooperation and cohesion. Governments should be accountable to their people, and all society's institutions – corporations, organizations and groups including us – should be accountable for the impact of their actions" (Oxfam 2017a).

The creation of a worldwide influencing network (including the further development of communication, advocacy and digital strategy) is one of the key priorities envisaged for this transformational shift. Oxfam's work on global campaigning – whether on climate change, displacement or sustainable development – makes use of celebrities to disseminate messages worldwide with all the implications already discussed for Save the Children in this regard.

Oxfam International is committed to both humanitarian and development work, declaring that "as well as becoming a world leader in the delivery of emergency relief, Oxfam International implements long-term development programs in vulnerable communities" (Oxfam International 2018b). Oxfam does not only define itself as a development organization that is also working in humanitarian response but considers humanitarian action as a "work that integrates life-saving response with building resilience" (Oxfam 2013, 11). The link between humanitarian action, human security and development is explicitly stated in the Strategic Plan goal 3 "Saving lives, now and in the future", where the specific objectives include, beside the provision of humanitarian assistance, resilience, institutional capacity building for human rights protection and women's rights.

In order to achieve its goal, the organization applies a "combination of rights-based sustainable development programs, public education, campaigns, advocacy, and humanitarian assistance in disasters and conflicts" (Oxfam 2017a). The rights-based approach is among the organization's guiding principles (Oxfam 2013). Within Oxfam, human rights are perceived as an operational concept: respect for human rights is t considered a prerequisite in the fight against poverty and inequality. Oxfam asserts that it has operationalized the human rights approach into its work around livelihoods, access to basic services, protection, the right to be heard, and inclusion. As with Save the Children, the combination of humanitarianism with development and the embracement of a rights-based approach indicate the role that Oxfam wants to play in global governance.

There are a further two other elements which the organization considers fundamental and characteristic of its action: making people's voices heard and empowering them (especially women). On its website, Oxfam International represents this two-fold approach to fighting poverty by two icons. The first symbol (a megaphone) refers to the ability of people to claim their own rights: "When people have the power to claim their basic human rights, they can escape poverty – permanently. This core belief underpins our development programs in more than 90 countries. With our partners, allies and with local communities, we help people to claim rights for themselves" (Oxfam International 2018c). Expressions referring to the importance of making people voices heard to affect social change are consistently repeated throughout Oxfam's strategic documents. "Oxfam's Strategic Plan to 2019 has a vision that sets local communities and the voices of women, men and young people at the centre of change" (Oxfam 2013, 5); "we will put a particular focus on gender justice and empowering poor people to make their voices heard"' (Oxfam 2013, 8); "Oxfam believes that people living in poverty who claim their rights and make their voices heard constitute an enormous source of hope for real change and greater power in people's lives" (Oxfam 2013, 9); "Bring the voices of poor and vulnerable communities into debates about development and prosperity" (Oxfam 2013, 19). The second icon, the stylized figure of a woman with a speech bubble, indicates women empowerment, which symbolically is, again, represented by the ability to speak out: "Human development is driven by empowered women. But women and girls are still massively underrepresented and often oppressed. We work to help them speak out and demand justice, and to assert their leadership." (Oxfam International 2018c).

The empowerment of people and giving them voice is fundamental to Oxfam's public identity (as it is with many other NGOs) and was frequently mentioned by my interviewees. This is an important point that will emerge again in the analysis of NGOs' communication material. It is important to note that the empowerment of people through their own voices is far from unproblematic. By focusing on an Oxfam project developed exactly for this purpose, entitled *Listening to the Displaced*, Rajaram (2002) has pointed out how the organization's representation did not manage to distance itself from more traditional depiction of refugees "in terms of helplessness and loss". Far from allowing displaced people to speak for themselves and account for their individual experience of displacement, the project, rather, reproduced an image of refugees in de-politicized and de-historicized terms. While acknowledging Oxfam's admirable intention to challenge the systematic silencing of refugee identity, the scholar noticed how the "particular bureaucratized knowledge about refugees and the methodology for "listening' to them" (Rajaram 2002, 248) concur to a sense of objectivity and distance that conceal the author as mediator in the publication of refugee experiences. The refugees' voices, included in the project to understand displaced people' personal perspective and empower them, are not only reduced to those of humanitarian subjects, but they are also denied the right to account for their experience in politically relevant terms. In the new humanitarian rhetoric around making people's voices heard, of which Oxfam is a prime exponent, Western systems of meaning are thus reproduced. (Rajaram 2002).

For Oxfam advocacy plays a fundamental role in the organization's aspiration to play a major role in changing the world. The influencing component of Oxfam's strategic programme is seen as part of its strategic objective as much as development and humanitarian response. Influencing is judged to be crucial to the achievement of their long term development goals and humanitarian achievements. According to an Oxfam advisor interviewed for this study: "we are knocking at every door to influence what we are campaigning for: different parties, ministries, donors, Heads of States, despite their political colour. At the end of the day, it is always about rights. If politics helps the cause for rights, we will work with it. Of course, we do a lot of context analysis, power analysis, stakeholder analysis. But, for instance, also in the case the European Migration Campaign and the political position of EU to close its borders, of course we take money there – even though EU gives the money in the neighboring countries to keep the refugees there. We anyhow take the money to address the needs. Oxfam is big, we can still work on the humanitarian side and the advocacy side at the same time. Then, of course, different Oxfam offices may decide to take money from some government because they are trying to influence it and they want to remain independent. For instance, Oxfam USA does not accept USAID money (but this is linked to advocacy on domestic issues)."[5] This quotation clearly shows how the intention to be operative on the ground and able to influence decision makers helps determine the extent to which Oxfam interacts with politics, donors and their policy agendas.

The possibility of overcoming political differences is based on the focus on rights, considered as something neutral, distinct from politics, a sort of technical field to which Oxfam can appeal to in order to avoid open reference to power and politics. The statement also underlines the crucial way in which Oxfam perceives its role in global governance: as an organization 'big enough' to be able to implement humanitarian work (portrayed as exogenous to the political realm) with advocacy (considered as openly political).

In order to achieve transformational change, over the last decades, Oxfam has developed strategies that complicate even further the already complex relationship that the organization has with power. On the one side, Fassin has pointed out how

---

[5] Interview with Oxfam Policy Advisor, 8th March 2017, Beirut.

Oxfam is the example of what he defines a "paragovernmental agency"'. As he explains, "humanitarian intervention has become a policy of nation states, whether because governments are developing their own activity in the field (France, for example, has had several ministers for humanitarian aid) or because they delegate it to paragovernmental agencies (such as the Oxford Committee for Famine Relief [Oxfam] in the United Kingdom)" (Fassin 2007, 508). Oxfam, like other major NGOs has been increasingly working in partnership with private capital. As Barnett and Weiss (2008) have argued, the problematic aspect of this collaboration revolves around the values and practices that dominate the social mechanism of the market, where everything is given a price. Therefore, they ask, how can humanitarian practitioners accommodate humanitarian principles with the calculations of the market? In both cases, it is evident how the NGO works in a complicated environment where there are political and economic interests at stake that eventually affect, in one way or another, not only the aspirations of the organization, but also its practical work on the ground role in the international community. As previously discussed in relation to SCF, dependence on government funds and collaboration with private capital are not without consequences. Whereas it appears quite straightforward that national interests, private capital and humanitarian action are inspired by quite diverse logics, their strict interrelation raises also questions regarding the NGO's space of manoeuvre in carrying out activities that may present a conflict of interests, an apparent threat to agency emphasis on maintaining independence. While Fassin's definition of Oxfam as paragovernmental may seem exaggerated, the implications of depending on institutional funding and working in collaboration with private capital cannot be underestimated.

4.3.2 Oxfam and Syrian Displacement

Oxfam works with people whose lives have been affected by natural or human-induced disasters around the world. Thus Oxfam is active in many displacement contexts, primarily providing water, food, sanitation and protection services. With regard to those affected by the Syrian war and displacement, the NGO operates in Greece, Italy, Jordan, Lebanon, (North) Macedonia, Serbia and Syria. Inside Syria, Oxfam has been working on the rehabilitation of water infrastructure, provision of clean water, public health promotion, solid waste management and supporting livelihoods. In Syria's neighboring countries, Jordan and Lebanon, the organization's intervention has been focused on water, sanitation, cash assistance and service information/referral projects to advise people about their rights and refer them to the most adequate support services. Along the migration route, Oxfam has worked with newly-arrived people by providing assistance (water and sanitation facilities) and protection (psychosocial support and legal counselling) in Serbia and (North) Macedonia. In Italy and Greece, the organization has distributed food and non-food items and provided psychological and legal support to those in need.

Alongside this emergency response the NGO works to influence policy makers to promote the changes needed for the improvement of refugees' lives Influencing plays a pivotal role also in its emergency work. In Oxfam's words: "Providing life-saving support to the millions of people affected by this devastating conflict is essential but it is not enough. We have been campaigning and advocating for an end to the fighting, and a sustainable and inclusive political solution since the beginning of the crisis. The continued violence, bloodshed and suffering in Syria represents a catastrophic failure by the international community to bring peace and security. We will continue to call on all parties to the conflict to commit to ending the massive violations, stop any arms transfers and guarantee humanitarian access and protection of civilians, whether inside Syria or in neighbouring countries" (Oxfam International 2018a).

The organization also carried out several social communication campaigns such as *Protect the lives of refugees and migrants* (Oxfam 2016b), or *Join hands #with-Syria* (Oxfam 2016a), through which it asked people around the world to share selfies with the dove symbol to indicate solidarity with Syria. As part of the more overtly political *Syria Refugee Crisis: Is your country doing its fair share?* (Oxfam 2016c), the agency urged EU member states to do more in terms of resettlement.

The analysis of the work that Oxfam carries out in response to Syrian displacement has highlighted that the organization has been mainly focusing on two streams of action. The first, based on relief assistance in line with traditional emergency support (e.g., provision of basic services) has also been combined with psychosocial support within the more holistic perspective of the rights-based approach. The NGO has sought to realise aspirations in its programmatic documents for a comprehensive and sustainable response to human suffering. This approach underlines Oxfam understanding of its role within the global governance regime for displaced people and that it should not be confined to provision of water and building of latrines, but involve active engagement in promoting and protecting peoples' rights. This stream of action is more explicitly political. It openly speaks to state politics and is embedded in Oxfam's stated commitments to the humanitarian principle of neutrality and need to speak out on human rights violations.

4.4 CARE

The *Cooperative for American Remittances to Europe* (CARE) was founded in 1946 by 22 US-based charity organizations with civic, labour, and faith-based backgrounds. Like SCF and Oxfam, its initial scope was to provide food assistance to people suffering famine in the aftermath of World War II. Once post-war Europe was on the road to recovery CARE became operational in Asia, and subsequently in Latin America and Africa. In addition to food relief, CARE started to provide medical support and non-food-item assistance to people affected by conflicts. To represent the wider geographical extent of its operations, it changed its acronym to the *Cooperative for American Relief to Everywhere*. By the 1970s it had expanded its areas of intervention from immediate assistance to long-term programs of recovery and rehabilitation. Like many other NGOs it became increasingly focused on female inequality in the 1980s and today has an explicit organizational focus on gender. There was a further re-branding in 1993 when its acronym was again changed to the *Cooperative for Assistance and Relief to Everywhere*. Today CARE International is a confederation of 14 member organizations (with seven candidate and affiliate off-shoots). It works in over a 100 countries, mainly focused on food security, education, health and community wellbeing, women's economic empowerment, water and sanitation.

With regards to humanitarian principles CARE defines itself as an agency which is "Independent of political, commercial, military, ethnic or religious objectives CARE promotes the protection of humanitarian space. We provide assistance on the basis of need, regardless of race, creed or nationality addressing the rights of vulnerable groups, particularly women and girls" (CARE 2017). CARE, like SCF and Oxfam, similarly adheres to the international guidelines and statements of principle mentioned above. CARE respects a set of internal Programming Principles which include promoting empowerment, working in partnership with others, ensuring accountability and promoting responsibility, addressing discrimination, promoting non-violent resolution of conflicts, and seeking sustainable results.

CARE International aggregated income amounts to 725,640 million euro. Sixty four per cent of revenue in 2016 came from government and institutional grants, other donor contributions (27%), in kind (6.5%) and other income (2%) (Care International 2017). Expenses are divided among development activities (57%), humanitarian activities (28%), and supporting services (14%) (Care International 2017).

4.4.1 CARE and Society

CARE's work, combining development and humanitarian assistance programmes, aims at ending poverty so that people can "live in dignity and security" (CARE 2017). CARE International's Programme Strategy focuses on reducing poverty, understood as the result of power imbalances and social injustice (CARE 2014). The organization considers gender inequality and humanitarian crisis as two main factors in producing injustice and insecurity in the world. CARE's humanitarian action is combined with development and resilience-building objectives to achieve lasting impact. Specifically, the organization humanitarian goal for 2020 is to 'to continue to strengthen our humanitarian work to have a lasting impact for people affected by humanitarian crises, with a special focus on women and girls' (CARE 2016a, 2). As Barnett has noted, CARE, like SCF and Oxfam "began as relief organizations and thus initially saw themselves as part of the humanitarian system, once relief was no longer a priority they soon tackled reconstruction and development activities and no longer identified as closely with a "humanitarian" system still very much defined by relief" (Barnett and Weiss 2008, 5).

The role that the organization acknowledges for itself in global governance is confirmed by several statements disseminated in official documents, such as: "CARE recognizes it must continue to lead and contribute towards guaranteeing that globally humanitarianism plays an essential role in overcoming social injustices and inequality"' (Care International 2016, ii). In order to achieve its mission, the NGO works in collaboration with the private sector, and it proudly presents the corporate partners and commercial brand with whom it collaborates (CARE 2018b).

In all its programs, both in development and humanitarian settings, CARE has a focus on women and girls. According to the organization, poverty and discrimination affect women disproportionally and therefore the NGO has decided to pay particular attention to gender empowerment. For CARE the gender approach is intended as a particular attention to the needs of women and girls rather than in its wider meaning, as an approach sensitive to any gender difference. In CARE's words: "globally, social injustice and gender inequality leave women highly vulnerable and disempowered to protect themselves during disasters and to reduce their risk to future disasters. Women and girls struggle to be heard and to be given the opportunity to make decisions that could save themselves and their families and communities during disasters." (Care International 2016, iii). In practice, this translates into long-term sustainable development programming to "create sustainable transformative changes in gender equality (…) CARE therefore strives to embed gendered disaster responses, disaster risk reduction (DRR), resilience and climate change adaptation as part of its long-term programming to address social injustices, weak governance and vulnerability. Failure to do this can undermine any impact achieved through humanitarian or development programming, or worse yet lead to 'double disasters' especially in terms of GBV [Gender Based Violence] against women and girls during crises and in the transition from relief to recovery due to the possibility of socially disrupted or destroyed households and communities" (Care International 2016, 16).

CARE's focus on gender is relevant when trying to grasp the role that the organization wants to have in global governance. There are a wide range of approaches that look at the production and reinforcement of gender inequality and attempts to challenge them. According to the United Nation agency dedicated to gender equality and the empowerment of women (UN Women) the different approaches can be situated along a spectrum that goes from *gender neutral*, an approach that does not reinforce existing gender inequalities, passing through *gender sensitive,* programs aimed at redressing existing gender inequalities, to *gender transformative*, approaches that attempt to re-define women and men's gender roles and relations. (UN Women 2018). In this sense CARE has definitely embraced the gender transformative model, "transforming the power dynamics and structures that serve to reinforce gendered inequalities" (Hillendrand et al. 2015).

Such contemporary emphasis on women's rights has been criticized as a neocolonialist and culturalist approach that considers the Western perspective as the universal standard to which other societies should adhere for their own good (see among others, Abu-Lughod 2002; Mahmood 2011; Grande 2016). Grande has argued that humanitarian discourse on women rights" functions as a powerful agent of social individualization" (Grande 2016, 78), whereby only women' individual rights are promoted and protected – in line with the Western understanding of human rights – while collective rights are dismissed as not really in the interests of women and always assumed to have been socially imposed on them. In this cultural reading, humanitarian organizations intervene to transform power dynamics that do not match with their own cultural background. Besides the obvious neo-colonialist and culturalist implications of such an approach, the problem, insightfully pointed out by Grande, is that women rights discourse advances a message strongly based on individualization rooted in neoliberal logic one which structures and discourses based on collective rights.

The gender transformative perspective is part of the wider rights-based approach embraced by the organization and explicitly mentioned within the principles of the organization (CARE 2017). The attention to rights is pivotal in two of the three main core elements of CARE's approach: *Strengthening Gender Equality and Women's Voice* and *Inclusive Governance.* Within the framework of the latter, the organization works on women empowerment and transform unequal power relationships, 'based on CARE's commitment to the rights of all people to live free from poverty'. The NGO places 'particular emphasis on strengthening the voice of women and girls and enabling them to influence the decisions that affect their lives. We support women's struggles to achieve their full and equal human rights, including girls' right to education. This includes balancing practical, daily, individual achievements with long-term efforts to challenge unequal social rules and institutions' (CARE 2018a). Beside the focus on women empowerment that I have already discussed in the previous paragraph, it is clear from these kinds of affirmations how the rights-based approach is entangled with the aspirations the NGO has in terms of society transformation and its role in global governance. Under the second element of

CARE seeks to play an active role in global governance, arguing that "CARE promotes inclusive governance in three key ways; by empowering poor and excluded people to know and act on their rights and represent their interests; by influencing those in power, such as governments, traditional leaders in the private sector, to be more responsible, responsive and accountable; and by brokering linkages and convening spaces which enable effective and inclusive relations and negotiation between the two" (CARE 2018a).

My interviews indicated that the complex relationship that the organization has with politics and global governance is not unknown to its staff, who, on the contrary is very aware of the risks and the opportunities implied in this complicated relationship. As a CARE communication and advocacy staffer affirmed: "let's start by stating that CARE is apolitical. Of course, the situation on the ground is more complex. And we must also say that all communication is political. For example, take the case of the safe zones in Syria and the debate in Lebanon around the Syrian refugee return to those zones. CARE would publicly say that they do not go into politics, that UNHCR is dealing with these kinds of issues, that the war at the moment is not over. They would always make the point. At the same time, since this is a very sensitive issue in Lebanon, and the Lebanese population has been very welcoming with the refugees, CARE would not say that they are against safe zones because Lebanese public opinion would be very against such a position (because, let's be clear, they want the Syrians out). So, in order to avoid politics, CARE would take a human rights approach, saying that if there is a legitimate authority (such as the UN) that guarantees the safety and security of those zones and the fact that Syrians are moving to those zones under no coercion or incentives, in that case, we would support the idea. (…) Politics goes into that whether we like it or not. But we take a human rights perspective, we do not talk about politics."[6]

Once again, the realm of human rights is considered a neutral and technical dimension that the organization can appeal to in order to avoid interconnections with politics. Interestingly, while this account confirms high level of staff reflexivity around the complexity and dilemmas of the relationship between humanitarianism and politics, it seems at the same time to disregard that human rights discourse is intrinsically political.

This is important especially in relation with the literature that has offered a classification of the major humanitarian organizations according to their relationship with politics (see *Chap. 2*). In all different studies, CARE is always one of the organizations positioned furthest from the 'complete-independence-from-politics' end of the spectrum. Stoddard has classified CARE as typically Wilsonian (Stoddard 2003, 28). Similarly, Barnett (2009) has used CARE as a paradigmatic example of an *alchemic* and *dependent* organization, one that aspires to address the root causes of suffering while being dependent on State funding and agendas. Alchemic organizations like CARE rebuff accusations of being too close to States by framing political interventions in technical terms Analysis of CARE documents and interviews with CARE staffers confirmed this.

Interestingly, however, for some CARE staff the role of the organization vis a vis global governance depends, at least to a certain extent, on the level of their office, whether national, regional or international. As a CARE emergency coordinator told me when asked how she understood CARE's role in global governance: "I am not sure how to answer this question because the field is the last level of CARE organization. Talking about global governance you should talk to the people who work at the international level." Similarly, discussing the complicated relationship of humanitarian action with politics, she noted that: "for CARE Greece it is not a very big issue. It is different for national and international NGOs because in our case any advocacy product that could have any political implication goes through CARE Geneva. At the national level, our work is not affected by politics or the political situation because our advocacy anyhow is not targeting the Greek government but EU members."[7]

Although highly aware of the key role that the NGO plays in the international arena, indeed taking it for granted, what the NGO staff I talked to suggested that humanitarian action had political implications at the highest international level,

---

[6] Interview with CARE Communication and Advocacy Manager, 7th March 2017, Beirut.

[7] Interview with CARE Emergency Coordination Officer, 13th June 2017 via Skype.

while at the local level, it constituted an intervention only related to providing aid, something seen as disconnected from politics and any particular role in global governance.

4.4.2   CARE and Syrian Displacement

CARE International works with displaced Syrians and host communities in Syria, Jordan, Lebanon, Egypt, Turkey, Greece and the Balkans. In Syria CARE has provided life-saving assistance and education services, distributing food and non-food items (NFIs) while building capacities of local organizations providing assistance on the ground. In host states CARE has focused on protection, distributing relief items, and water and sanitation. It has provided psycho-social support to women, men and children to help them cope with their experience of trauma, loss, and gender-based violence (GBV). CARE has provided information on how refugees can access health, legal and social support, and organized raising awareness sessions on GBV, child protection, and early marriage. Protection services have been provided in Jordan, Egypt and to a lesser extent in Turkey. CARE has also distributed emergency cash assistance (in Jordan, Lebanon, Egypt, and Turkey), hygiene kits and dignity kits for women and older people (Lebanon and Turkey), food (Turkey), non-food items (Turkey, and Lebanon), and water and sanitation (Lebanon and Turkey). In the Balkans and in Greece, CARE has distributed food and NFIs and supported the establishment of centres where information and translation services are provided by volunteers.

It should be noted that CARE's gender transformative programming has consequence for the role that the organization seeks to play in global governance. Far from wanting to contest the importance of gender sensitiveness on a general level, it is rather important considering the social implications of an approach based on such specific socio-cultural premises completely different from the area of intervention. In other words, the political consequences of a self-proclaimed technical and universal rights-based approach of a Western-based organization (with a transformative agenda) into the social dynamics of a Middle Eastern context needs to be acknowledged and understood into the framework of the role that CARE plays, and intends to play into global governance.

CARE appears to have been the only of the four studied NGOs to have set up specific Twitter (CARE 2012) and Facebook (CARE 2016b) pages around its response to the Syrian crisis.

Through social media CARE has launched such campaigns as *Dear World – We are People* ('Syria Crisis: Dear World – We Are People' 2016) and *What Would you Take?* (CARE 2016c) with the hash tag #Syria2000days. CARE's advocacy conforms with what Chouliaraki has defined "post-humanitarian communication"(Chouliaraki 2010), a strategy that has abandoned the focus on pity and grand-emotions for a more playful engagement that makes the spectator's emotions the measure of the understanding of the suffering of others.

4.5   Médecins Sans Frontières

Médecins Sans Frontières (MSF), was founded in 1971 in France by a group of volunteers, including doctors and journalists Among its 14 founders the most prominent was Bernard Kouchner, a French doctor, who at the time of the Biafra crisis, was working for the ICRC, but resigned in protest at the ICRC's failure to speak out against Nigerian government tactics to starve Biafran secessionists into submission. For Kouchner, ICRC and its staff were "accomplices in the systematic massacre of a population" (Kouchner cited in Chandler 2001, 684). Chandler (2001) has argued that the rise of MSF to global prominence has introduced two new humanitarian principles. The first has to do with the principle of denunciation that is today one of the agency's fundamental principles. The concept was so defined by James Orbinsky when MSF was awarded the Nobel Peace Prize in 1999: "silence has long been confused with neutrality and has been presented as a necessary condition for humanitarian action. From its beginning MSF was created in opposition to this assumption. We are not sure that words can always save lives, but we know that silence can certainly kill". (Orbinsky cited in Chandler 2001, 685). The second principle, and one that was subsequently confirmed by NGOs going into Iraq, Myanmar and other contexts in defiance of State prohibitions on access, is the right to intervene, disregarding state sovereignty in the name of higher principles of humanity

MSF may have emerged in opposition to what Kouchner regarded as ICRC's passivity but, nevertheless, MSF is firmly rooted within the Dunantist tradition – based on the principle of independence of humanitarian action (Tong 2004). MSF has stated that its "actions are guided by medical ethics and the principles of impartiality, independence and neutrality" (MSF 2018a) and "claims full and unhindered freedom in the exercise of its functions. (MSF 2017b)". As Tong (2004) has pointed out, these principles are understood in a pragmatic way, that takes account of contextual complexities in which humanitarian aid is needed and delivered MSF assets that to respond to the humanitarian imperative it may be necessary to privilege the principles of impartiality or independence over that of neutrality. In other cases, the organization can decide not to intervene or withdraw from an emergency when there is the possibility that aid is contributing to the perpetration of violence.

Although MSF has signed the Code of Conduct for the International Red Cross and red Crescent Movement and NGOs in Disaster Relief, it has withdrawn from Sphere fearing the project overemphasizes quality and accountability indicator to the detriment of considerations around political contexts and power dynamics (Tong 2004).

By distancing itself from Sphere and the UN Cluster System MSF may act as if it is a humanitarian outsider, but it has become a powerful and well-funded movement. Now comprising 25 sections (recently renamed by MSF as associations), the agency is active in over 80 countries across the world. Given the exclusive focus of the organization on humanitarian activities, the 80% of the total income of 1,5 billion Euro devoted to relief activities. As opposed to the other organizations selected for this study, and indeed almost all other international NGOs, MSF overwhelmingly depends on private individuals donating small sums, 95% of its funding in 2017. Since 2016 MSF no longer accepts funding coming from any EU member States and Institution in protest against European migration policies (MSF 2016h). MSF identity is rooted in its assertion that the agency "remains fiercely independent of both governments and institutions" (MSF 2018c). The strategy of relying on non-state funding from multiple supporters has consequences, requiring a consistent and resource consuming attention to fundraising and maintaining contact with myriad supporters.

4.5.1 MSF and Society

Unlike the three other NGOs studied here, MSF focuses only on medical humanitarian assistance. The organization intervenes in situations of armed conflict, natural disaster, endemic and epidemic diseases, social violence, and healthcare exclusion. It offers basic healthcare, carries out vaccination campaigns, performs surgery, fights endemics, rehabilitates and runs hospitals and clinics, operates nutrition centres, and provides mental healthcare. Where necessary the organization sets up sanitation systems, supplies safe drinking water, and distributes relief (MSF 2018d). Because it has traditionally focused on humanitarian assistance MSF has been classified as" single-mandate" organization, against" multi-mandate" agencies combining humanitarianism with development (Tong 2004). As Tong has observed, this does not mean a complete dismissal of a long-term perspective, but rather its sidelining in the name of immediate needs.

Although scholars (Chandler 2001; Barnett 2009) have defined the NGO as a rights- based organization, they have also later noticed how after the intervention in Afghanistan in the 2000s, MSF changed direction refusing to participate to reconstruction programmes and going back to a need-based approach[8] (Barnett 2009). MSF's preference is confirmed by the paper that MSF presented at the World Conference of Humanitarian Studies (DuBois 2007) that outlines the organization's criticism of the right-based approach. It is further indicated by MSF's statement that "We offer assistance to people based on need. It doesn't matter which country they are from, which religion they belong to, or what their political affiliations are. We give priority to those in the most serious and immediate danger"' (MSF 2018a). A needs-based approach underpins its work, endeavouring whenever possible, to close operations as soon as life-saving needs are over. As explained by MSF, "the closure of a programme reflects MSF's specific mandate to provide medical humanitarian assistance. The decision to withdraw is based on our experience and analysis of the

---

[8] As Barnett (2009) illustrated, humanitarian organizations active in Afghanistan become increasingly involved in governance activities, promoting democracy and human rights and connecting their relief work with the reconstruction operations. Although MSF had already included long-term activities within its humanitarian work, the organization refused to participate in reconstruction programmes on the basis that they were incompatible with their humanitarian mandate.

situation, and the imperative to make choices so that we can devote our assistance to the most vital needs" (MSF 2018b).

For MSF, speaking out or *témoignage* (bearing witness) is a fundamental aspect of the agency's identity. and has become one of the distinctive characteristics of the organization since its origin. This is intended as "an effort to bring a forgotten crisis to public attention, to alert the public to abuses occurring beyond the headlines, to criticise the inadequacies of the aid system, or to challenge the diversion of humanitarian aid for political interests" (MSF 2017a). While MSF has firmly affirmed its neutrality and impartiality Slim has classified the agency in terms of "active impartiality'" (Slim 1997, 349). According to the author, MSF's specific interpretation of the principle of impartiality does not have to do with the various parties involved in a conflict, toward which the organization will always maintain its impartiality, but has rather to do with the actions of the various factions. In cases the NGO observes human-rights abuses, it will publicly speak out to denounce or condemn the violation.

Also, advocacy and campaigning play an important role in the organization. Similarly, to the other three NGOs, MSF dedicates a crucial attention to campaign activities. The organization's campaigns have mainly focused on specific health issues, such as medicine access (MSF 2016a), and drugs prices and patents (MSF 2016b, f), or more related to broader humanitarian issues such as the campaign #stayingalive (MSF 2016g) and #notatarget (MSF 2016e).

As noted, a further distinctive feature of MSF work is that it does not participate in the UN cluster system. MSF objects to the emphasis on coordination as a humanitarian goal in itself. For MSF, coordination has to be "useful, guided by the reality of the situation on the ground and directed toward concrete action. The limits of the mechanisms of coordination must be recognised: faced with political interference in humanitarian aid, the solution resides not in multiplication or strengthening of technical measures, but rather in the need for humanitarian organisations and the international community to highlight political constraints and ascribe responsibilities" (Dubuet and Tronc 2006, 2).

MSF's views on the complex interaction between humanitarian action and politics are markedly different from the other three NGOs here studied. According to a former MSF head "humanitarian action is noble when coupled with political action and justice. Without them is doomed to failure" (Destexhe cited in Chandler 2002, 43). This vision was widely reflected in comments from MSF field staffers I interviewed at the time of the controversial EU-Turkey deal mentioned earlier. An MSF communications manager told me "this is a political crisis and has been created by the European Union and other member states. To respond to this crisis, we have to take a political stance."[9] Another MSF interviewee explained that: "Sometimes, we would like to continue with public advocacy but because we are also carrying out private negotiations with the army and the government, it is better to not publicly speak out, because the medical needs are still there, and we need to address them as

---

[9] Interview with MSF Communication Manager, April, 25th 2017, Athens.

soon as we will have the opportunity."[10] MSF has not hesitated to withdraw from operational engagement in the belief that it,could do more harm than good (Chandler 2002) or completely refused to intervene (Fassin 2007). Interestingly, when I directly asked a senior MSF manager about MSF's involvement in governance, I was told that "we do not participate in global governance. We have been very involved as a global health actor, when the other were not doing anything but we do not have a role in global governance."[11] For MSF' position toward politics, MSF has been considered as part of a 'Latin' tradition of humanitarianism that, in opposition to the Anglo-Saxon model, tends to have an antagonistic relationship with power (Tong 2004). Rieff has defined this approach as "disobedient humanitarianism" (Rieff 2003, 272).

4.5.2 MSF and Syrian Displacement

MSF has responded to the Syrian emergency in Bulgaria, France, Greece, Italy, Iraq, Jordan, Lebanon, Syria, Turkey and Tunisia. MSF's Syrian crisis response has focused exclusively on humanitarian assistance delivered on the basis of needs and has not included any form of development activity. Unlike other agencies investigated who declined to provide details of in-country operations citing security concerns. MSF has been quite vocal, declaring that:

> the Syrian government has not granted MSF authorisation to operate in the country, despite repeated requests, and insecurity has limited MSF's ability to provide assistance in opposition-controlled areas. Following the Islamic State (IS) group's abduction and release of MSF staff in 2014, and the impossibility of obtaining the necessary safety guarantees from its leadership, MSF withdrew from IS-controlled areas. In 2016, MSF continued to operate directly in six medical facilities in regions controlled by other opposition forces across northern Syria and provide distance support to Syrian medical networks in areas where MSF cannot be directly present. (MSF 2016d).

Inside Syria MSF has mainly focused on provision of medical supplies, basic health services and surgery, maternal health, vaccination campaigns, mental health and psychosocial support. In some areas it has also implemented water and sanitation programme and relief items distributions.

In refugee hosting states MSF has offered primary health care, psychosocial support, maternal healthcare, mental health, health promotion, and (in Jordan) medical evacuation of the wounded). In Tunisia, it has supported local organizations building the capacities of Tunisian and Libyan fishermen and coastguards in SAR and management of corpses. Along the migration route, MSF has provided mental health care in reception centres (in Italy), and medical care, shelter, relief items and hygiene kits at borders. In Greece, after the EU-Turkey deal left migrants stranded in Greece, the organization has provided lifesaving surgery and medical care to

---

[10] Interview with MSF Humanitarian Affairs Officer, May 17th 2017, Amman.

[11] Interview with MSF Director of Operational Support, April 25th 2017, Athens.

people on the move.. In March 2016, MSF withdrew from the Moria detention facility on the Greek island of Lesbos, stating that it had become "a pre-removal detention centre, offering little guarantee of respect for human rights'(MSF 2016c).

With regards to the advocacy around Syrian displacement and Mediterranean migration, MSF has produced several reports, statements and press releases, severely rebuked governments and used social media to confront state policies. Overall, the NGO's communication appears quite different from the dominant Chouliaraki's post-humanitarian communication (Chouliaraki 2010) based on grand emotions and in search for (individual) emotive engagement with the humanitarian subject. MSF communication results more openly political both in the ways humanitarian crisis are presented to the public (for information and fundraising purposes), and for its often confrontational position toward governments' politics.

In relation to SAR operations which it began in 2015 MSF has operated vessels crewed by MSF staff, and has provided medical personnel to other boats working to save lives at sea. Cuttitta has shown how MSF's SAR work has from the outset been presented as highly political, in contrast with the position taken by other NGOs supporting maritime SAR. For MSF, "humanitarian work can't be kept separated from the investigation and critique of the causes that make humanitarian work necessary" (Cuttitta 2017, 9). Beside the primary goal of saving lives, MSF has publicised its SAR engagement and visibility in order to present "a more humanized image of migration, one which is alternative to the stereotypical picture of an invasion caused by criminal actors, and, second, to ask for safe passages" (MSF representative quoted in Cuttitta 2017, 10).

4.6   Conclusion

This chapter has shown the role that each of the organizations wants to have in global governance and how they position themselves in terms of its relationship with politics. Table 4.1 schematically outlines the four organizations' self-perception regarding their role in global governance and relationship with politics.

Save the Children perceives and presents itself as a key player in global governance as confirmed by the organization's complete embrace of the main characteristics of the new humanitarianism (multi-mandate mission, rights-based and transformative approaches, strong focus on advocacy). In contradiction however, Save the Children 's self-perception and self-representation seems based on a clearly apolitical stance, compliant with humanitarian principles and international standards. It consistently presents activities for the promotion and protection of children's rights as technical and apolitical. However, analysis has shown that through psychosocial and SAR operations, the agencies contributes to the SCF has demonstrated the effects of pathologization and depoliticization of refugees, indicating the highly political role that Save the Children plays in refugee governance.

Table 4.1 NGOs, global governance and politics

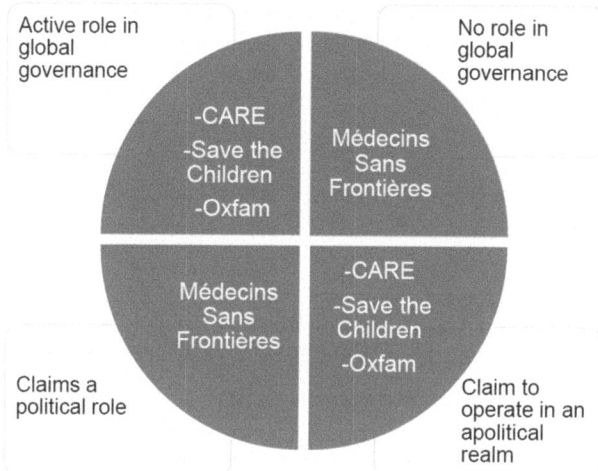

Oxfam shares with Save the Children a pre-occupation with forging a crucial role in global governance. It combines a strong relationship with governmental institutions with a more confrontational attitude toward state politics. Oxfam's focus on rights, people's voices and empowerment, as something neutral, unconnected from politics, is considered and presented as technical expertise, allowing Oxfam to avoid open reference to systems of power and politics. Despite the stress it puts on giving voice to beneficiaries, the analysis has shown how it consistently contributed to the depoliticization of the refugee issue.

CARE has also fully embraced and publicized the key role that it seeks to play in global governance and portrays itself as apolitical. Its gender transformative approach, through which the organization aims at redefining social relations and power dynamics, indicates its crucial ambiguity While urging a transformative approach it categorically denies any political will or political implications of its programs. CARE is aware of this tension and seeks to solve it by claiming to act only within the realm of human rights – which is considered by the organization as a neutral and technical space located outside the political sphere.

In complete contrast, MSF does not perceive and present itself as playing a role in global governance and has made its active engagement with politics a distinctive feature. MSF has vigorously insisted on its political (and financial) independence in order to be able to maintain space to maneuver and, when needed, to contest States. It has always considered politics an important aspect of its humanitarian intervention.

Overall, despite the four NGOs' different positions and self-representation regarding their role in global governance and relationship with politics, this chapter has shown how the work humanitarian agencies perform in the field is highly political and inevitably has a role both in humanitarian and wider global governance. Beyond each humanitarian NGO's self-positioning this study is interested in exploring specifically how their forms of visual representation contribute to shape the role that relief agencies perform in global, particularly refugee, governance. In the following chapters, the analysis will move to the investigation of the relief agencies visual communication to see how the visual practices of representation of the people they are assisting, especially Syrians on the move, relates with the roles transnational humanitarian NGOs want to have.

References

Abu-Lughod, L. (2002). Do Muslim women really need saving? Anthropological reflections on cultural relativism and its others. *American Anthropologist, 104*(3), 783–790.

Achilli, L.. (2016). Tariq Al-Euroba: Displacement trends of Syrian asylum seekers to the EU. 9290843721.

Barnett, M. (2009). Evolution without Progress? Humanitarianism in a World of Hurt. *International Organization, 63*(4), 621–663.

Barnett, M., & Weiss, T. G. (2008). *Humanitarianism in question: Politics, power, ethics*. Ithaca: Cornell University Press.

CARE. (2012). *Twitter CARE Syria response (@CARESyriaResp)*. Twitter 2012. https://twitter.com/CARESyriaResp

CARE. (2014). *Working for poverty reduction and social justice: The CARE 2020 Program Strategy*. http://insights.careinternational.org.uk/media/k2/attachments/CARE_2020_Program_Strategy-English.pdf

CARE. (2016a). *Care 2020 program strategy: Humanitarian response*. https://www.care-international.org/files/files/Summary-Humanitarian%20Response.pdf

CARE. (2016b). *Facebook CARE Syria response*. 2016. https://www.facebook.com/CARESyriaResponse/

CARE. (2016c). *What would you take: Syrian refugee crisis | CARE*. 2016. https://web.archive.org/web/20160528135836/http://www.care.org/emergencies/syria-crisis/care-for-refugees/what-would-you-take

CARE. (2017). *Vision and mission | Care international*. 2017. https://www.care-international.org/who-we-are/vision-and-mission

CARE. (2018a). *CARE's approach | Care international*. 2018. https://www.care-international.org/what-we-do/cares-approach

CARE. (2018b). *Corporate partnerships*. CARE. 2018. https://www.careinternational.org.uk/get-involved/corporate-partnerships

Care International. (2016). Empowering women and girld affected by crisis.

Care International. (2017). *Care international annual report FY16*. https://www.care-international.org/files/files/publications/Care_International_Annual_Report_2017_8921_ENG_12_Online(4).pdf

Chandler, D. (2001). The road to military humanitarianism: How the human rights NGOs shaped a new humanitarian agenda. *Human Rights Quarterly, 23*(3), 678–700.

Chandler, D. (2002). *From Kosovo to Kabul: Human rights and international intervention*. London: Pluto Press.

Chouliaraki, L. (2010). Post-humanitarianism: humanitarian communication beyond a politics of pity. *International Journal of Cultural Studies, 13*(2), 107–126.

Chouliaraki, L. (2012). The theatricality of humanitarianism: A critique of celebrity advocacy. *Communication and Critical/Cultural Studies, 9*(1), 1–21.

CHS. (2020). *Core humanitarian standard.* 2020. https://corehumanitarianstandard.org/the-standard

Cuttitta, P. (2017). Repoliticization through search and rescue? Humanitarian NGOs and migration management in the central mediterranean. *Geopolitics, 1–29.*

DuBois, M. (2007). Protection: The new humanitarian fig-leaf. *Dialogues, 4, 1.*

Dubuet, F., & Tronc, E. (2006). *United Nations: Deceptive humanitarian reforms?* MSF international activity report 2006.

Duvall, S. (2009). Dying for our sins: Christian salvation rhetoric in celebrity colonialism. In R. Clarke (Ed.), *Celebrity colonialism: Fame, power and representation in colonial and postcolonial cultures* (pp. 91–106). Newcastle: Cambridge Scholars.

Fassin, D. (2007). Humanitarianism as a politics of life. *Public Culture, 19*(3), 499–520.

Grande, E. (2016). "I'm doing it for myself!": The aggressive promotion of the individual self as the dark side of women's rights. In A. De Lauri (Ed.), *The politics of humanitarianims. Power, ideology and aid.* London: I.B. Tauris.

Hillendrand, E., Karim, N., Mohanraj, P., & Wu, D. (2015). Measuring gender-transformative change a review of literature and promising practices.

IFRC. (1994). Code of conduct for the international red cross and red crescent movement and Non-Governmental Organizations (N GO s) in disaster relief.

Kapoor, I. (2012). *Celebrity humanitarianism: the ideology of global charity.* Routledge.

Mahmood, S. (2011). *Politics of piety: The islamic revival and the feminist subject.* Princeton University Press.

MSF. (2016a). Access campaing. *Médecins Sans Frontières Access Campaign, 2016.* https://msfaccess.org/.

MSF. (2016b). *Access to medicines* | MSF medical response. Médecins Sans Frontières (MSF) International. 2016. /access-medicines

MSF. (2016c). *Greece.* Médecins Sans Frontières (MSF) International. 2016. http://www.msf.org/en/where-we-work/greece

MSF. (2016d). *International activity report 2016.* MSF – IAR. 2016. http://activityreport2016.msf.org/country/syria/

MSF. (2016e). *MSF – Not a target.* MSF – Not A Target | Médecins Sans Frontières. 2016. http://notatarget.msf.org/

MSF. (2016f). *Patent opposition database.* Médecins Sans Frontières (MSF) International. 2016. https://www.msf.org/patent-opposition-database

MSF. (2016g). *Stay alive.* 2016. http://stayingalive.msf.org/

MSF. (2016h). *MSF to no longer take funds from EU member states and institutions.* 16 June 2016. https://www.msf.org.uk/article/msf-no-longer-take-funds-eu-member-states-and-institutions

MSF. (2017a). *About MSF.* Médecins Sans Frontières (MSF) International. 2017. http://www.msf.org/en/about-msf

MSF. (2017b). *Who we are.* Médecins Sans Frontières (MSF) International. 2017. https://www.msf.org/who-we-are

MSF. (2018a). *About us.* Médecins Sans Frontières (MSF) International. 2018. https://www.msf.org/

MSF. (2018b). *Closing a Programme.* Médecins Sans Frontières (MSF) International. 2018. http://www.msf.org/en/closing-programme

MSF. (2018c). *MSF history.* Médecins Sans Frontières (MSF) International. 2018. http://www.msf.org/en/msf-history

MSF. (2018d). *Where we work.* Médecins Sans Frontières (MSF) International. 2018. http://www.msf.org/en/where-we-work-0

NGOs Platforms et al. (2017). *Stand and deliver. Urgent action needed on commitments made at the London conference one year on.* https://www.savethechildren.net/sites/default/files/libraries/Stand%20and%20Deliver%20digital.pdf

Oxfam. (2013). *The power of people against poverty.* Oxfam strategic plan, 2013 – 2019. https://www.oxfam.org/sites/www.oxfam.org/files/oxfam-strategic-plan-2013-2019.pdf

Oxfam. (2016a). *Join Hands #withSyria*. Twitter. 2016. https://twitter.com/Oxfam/status/709351389013663745

Oxfam. (2016b). *Protect refugees and migrants*. 2016. https://web.archive.org/web/20180627065149/https://actions.oxfam.org/international/

Oxfam. (2016c) *Syria refugee crisis: Is your country doing its fair share?* Oxfam International. 2016. https://www.oxfam.org/en/syria-refugee-crisis-your-country-doing-its-fair-share

Oxfam. (2017a). *Our purpose and beliefs I Oxfam International'*. 2017. https://www.oxfam.org/en/our-purpose-and-beliefs%20

Oxfam. (2017b). *Oxfam annual report 2015–2016*. https://www.oxfam.org/sites/www.oxfam.org/files/file_attachments/oxfam_annual_report_2015_-_2016_english_final_0.pdf

Oxfam International. (2013). Oxfam's role in humanitarian action.

Oxfam International. (2018a). *Crisis in Syria I Oxfam International*. 2018. https://www.oxfam.org/en/emergencies/crisis-syria

Oxfam International. (2018b). *History of Oxfam international I Oxfam International*. 2018. https://www.oxfam.org/en/countries/history-oxfam-international

Oxfam International. (2018c). *How we fight poverty I Oxfam International*. 2018. https://www.oxfam.org/en/explore/how-oxfam-fights-poverty

Oxfam International. (2018d). *Our work I Oxfam International*. 2018. https://www.oxfam.org/en/explore/issues-we-work-on

Pupavac, V. (2001a). Misanthropy without borders: The international children's rights regime. *Disasters, 25*(2), 95–112.

Pupavac, V. (2001b). Therapeutic governance: Psycho-social intervention and trauma risk management. *Disasters, 25*(4), 358–372.

Rajaram, P. K. (2002). Humanitarianism and representations of the refugee. *Journal of Refugee Studies, 15*(3), 247–264. https://doi.org/10.1093/jrs/15.3.247.

Rieff, D. (2003). *A bed for the night: Humanitarianism in crisis*. Simon and Schuster.

Save the Children. (2005). *Child rights programming handbook*. How to Apply Rights-Based Approaches to Programming. https://resourcecentre.savethechildren.net/node/2658/pdf/2658.pdf

Save the Children. (2014). *Most shocking second a day video*. 2014. https://www.youtube.com/watch?v=RBQ-IoHfimQ

Save the Children. (2016a). *Global corporate partnership brochure*. https://www.savethechildren.net/sites/default/files/Global%20Corporate%20Partnership%20Brochure.pdf

Save the Children. (2016b). *Save the children annual report 2015*. https://www.savethechildren.net/annualreview/ui/docs/Save_the_Children_Annual_Report_2016.pdf

Save the Children. (2016c). *Save the Children's Global Strategy: Ambition for Children 2030 and 2016–2018 Strategic Plan*. https://www.savethechildren.net/sites/default/files/Global%20Strategy%20-%20Ambition%20for%20Children%202030.pdf

Save the Children. (2016d). *Still the most shocking second a day*. https://www.youtube.com/watch?v=nKDgFCojiT8

Save the Children. (2017). *European refugee crisis*. Programme and Advocacy Report. https://www.savethechildren.net/sites/default/files/libraries/European%20refugee%20crisis_programme%20and%20advocacy%20report_Feb%202017.pdf

Save the Children. (2018a). *Advocacy*. Save the Children International. 2018. https://www.savethechildren.net/advocacy

Save the Children. (2018b). *Child rights governance*. Save the Children International 2018. https://www.savethechildren.net/what-we-do/child-rights-governance

Save the Children. (2018c). *Emergencies*. 2018. https://www.savethechildren.org.uk/how-you-can-help/emergencies

Save the Children. (2018d). *Minori migranti: il lancio oggi della campagna social "The Space Migrant"*. Save the Children Italia. 2018. https://www.savethechildren.it/press/minori-migranti-il-lancio-oggi-della-campagna-social-%E2%80%9C-space-migrant%E2%80%9D

Save the Children. (2018e). *Our humanitarian work*. Save the Children International. 2018. https://www.savethechildren.net/what-we-do/our-humanitarian-work

Save the Children. (2018f). *Our vision, mission and values*. Save the Children International. 2018. https://www.savethechildren.net/about-us/our-vision-mission-and-values

Save the Children. (2018g). *Syria every last child*. Every Last Child. 2018. https://campaigns.savethechildren.net/country/syria

Slim, H. (1997). Relief agencies and moral standing in war: Principles of humanity, neutrality, impartiality and solidarity. *Development in Practice, 7*(4), 342–352.

Sphere. (2020). *What are humanitarian standards?* | Sphere standards. Sphere. 2020. https://spherestandards.org/humanitarian-standards/

Stoddard, A. (2003). *Humanitarian NGOs: Challenges and trends*. 12. Humanitarian Policy Group (HPG) Briefing. HPG.

'Syria Crisis: Dear World – We Are People'. 2016. CARE. 2016. https://www.careinternational.org.uk/stories/syria-crisis-dear-world-%E2%80%93-we-are-people

Tong, J. (2004). Questionable accountability: MSF and sphere in 2003. *Disasters, 28*(2), 176–189.

UN Women. (2018). *Gender equality glossary*. https://trainingcentre.unwomen.org/mod/glossary/view.php?id=36&mode=letter&hook=G&sortkey=&sortorder=&fullsearch=0&page=-1

Van den Bulck, H. (2009). The white woman's burden: Media framing of celebrity transnational adoptions. In R. Clarke (Ed.), *Celebrity colonialism: Fame, power and representation in colonial and postcolonial cultures* (pp. 123–140). Newcastle: Cambridge Scholars.

Various NGOs. 2015. *Failing Syria. Assessing the impact of UNSCR in protecting and assisting civilians in Syria*. https://www.savethechildren.net/sites/default/files/libraries/FailingSyria_Report_March2015.pdf

Chapter 5
Threatening – The Refugee as a Threat

5.1 Introduction: The Refugee as a Threat

Contemporary media and public accounts have increasingly framed the refugee 'crisis' in terms of security, with refugees considered as masses to be managed and controlled, migrants pointed at with generic allegation of terrorist threat, and state borders closed and militarized. Securitization of migration may not be a new phenomenon (Saunders 2014) but it is one that has recently received a great deal of attention (see among others Bigo 2002; Pugh 2004; Huysmans and Squire 2009; Huysmans 2000; Musarò 2017; Vaughan-Williams 2015; Watson 2009). What all these scholars have in common is that they highlight different ways through which refugees are represented, described, and thought of as *threat*. Media and public accounts have consistently represented refugees through words such as plight, invasion, flood, hordes, or waves (Friese 2017). The "highly heterogeneous and (too) strongly mediation-dependent European politics created an array of – in most cases negative – interpretations of the Refugee Crisis" (Krzyżanowski et al. 2018). In line with this narrative, at the visual level, the images that have accompanied the news on refugees have mostly included overcrowded boats, long lines of people in need, and looming masses of people crammed at border fences.

One would expect the discourse of humanitarian organizations to challenge this view and propose an alternative, but this is only partly true. Humanitarian actors, and NGOs in particular, have tried to avoid the mainstream securitization framework and attempted to draw attention to the human, ethical and solidarity dimensions of the refugee issue. At the same time, humanitarian discourse has contributed to its securitization. This happens when refugees are represented as a *threat* and extraordinary measures (in this case humanitarian) are invoked to face the threat. "But threat to what?" one could ask. It is not possible to identify a unique subject at risk. According to the various securitizing representations, refugees can appear to constitute a risk for international order, legality, the status quo, territory and values.

© The Author(s) 2021
A. Massari, *Visual Securitization*, IMISCOE Research Series,
https://doi.org/10.1007/978-3-030-71143-6_5

There are different ways through which humanitarian discourse contributes to the representation of refugees as threatening. When humanitarian actors are perceived and perceive themselves as key player in the management of peoples' movements (Dillon and Reid 2000), there is an implicit assumption that this movement need to be managed and controlled. As scholars have pointed out, aid agencies are part of a security strategy to contain the disorder created by underdevelopment and wars (Duffield 2007). When NGOs contribute to the framing of an issue in terms of emergency, they automatically bolster calls for extraordinary measures. The exceptionality of the situation requires exceptional management to solve the state of emergency (Williams 2003). When refugees are represented as an indistinguishable group of people, their individuality, history, and context are erased not only from the picture in the literal meaning but also from our imaginary of the situation that has caused their displacement in the first place. Moreover, a certain kind of representation of the refugee issue as something drastically different from our own lives, or distant from our here and now ends up creating a 'we' the spectator of the crisis, and a 'they', the refugees, ultimately underpinning dynamics of inclusion and exclusion (Szörényi 2006).

Narratives like the ones just described are of course noticeable at the visual level. In front of a picture of lonely child staring mournfully at the camera (and therefore at us, the viewer) it is an understandable human reaction to think that something should be done, that that gaze requires some action from our side (Chouliaraki 2013). On the contrary, the kind of images that depict refugees as masses, that privilege groups of adult man over innocent children are much more likely to leave us uncertain about what should be expected from us as a reaction. As much as images of babies looking us in the eye can trigger an emotional response of empathy (Manzo 2008), images of overcrowded boats or masses of indistinguishable faces may arouse various degrees of discomfort, apprehension and fear (Bleiker et al. 2013). But the securitization goes also behind the emotional sphere. At the political level, the visual narrative of the threatening refugee contributes to reinforce a discourse on refugees that tends to depoliticize the causes of displacement and the agency of people on the move. In this simplified account, the refugee situation is presented as a question of safety, order, management and control, ultimately a question of security.

This chapter explores the different aesthetic patterns through which transnational humanitarian NGOs have contributed to a securitized representation of refugees, one in which people on the move are depicted as threat. Through analysis of five different visual themes – vectors without goal, visual rhetoric of emergency, iconography of migrants' boats, the use of conceptual structures and the visual construction of the 'other' – this chapter will show how relief agencies have depicted refugees in terms of menace.

5.2   Visual Threats

5.2.1   Vectors Without Goal (Fig. 5.1)

This picture has been published as 'photo of the week' on MFS's website. The image represents a long line of people with backpacks and bags walking, all in the same direction, through a rural area. In the language of visual social semiotics, the "visual syntactic pattern'" (Van Leeuwen and Jewitt 2001) of this picture is a narrative one. Its participants are represented in the process of doing something, there is an action going on (as opposed to the conceptual structure that shows the subjects in their essence, with no reference to a particular time and space). People's faces are not identifiable and the group at first glance seems mostly composed by men. Looking more closely we may discern a few children. However, individual details are not very relevant here since the subject of the photos – what is visual social semiotic, is called the "*represented participant*" of the image – is clearly the *group* of refugees. The effect is conveyed by portraying the subjects as fused in one line, with no separate distinctive characters. The hint about these people being refugees is given by the presence of luggage that functions as an index of their displacement.

Fig. 5.1 ©Alessandro Penso/MSF

The background of this image is also given importance because of its relative salience.[1] The lines of the trees, the grass, the mountains, and the clouds do not appear as a distant setting but constitute lines running parallel to the one made of refugees. This clearly visible and salient setting situates the action in a rural field, familiar scenery, a generic landscape to be found in many parts of Europe. Although the shape of the natural element elegantly accompanies the shape of the long line of refugees, this last element stands out in the image as the most salient and relevant. Not only is it the only not-natural element in the composition, but also the group of people presents a higher color contrast compared to the rest of the rather softer, almost pastel, colors of the landscape.

What I find most interesting in this picture lies in its narrative structure. Drawing on the literature that has gone behind the idea that the visual structure is a representation of reality, or that which has described it only in formalistic terms, Kress and Van Leeuwen (1996) have highlighted the importance of the visual structure in meaning production. Visual structures are ideological as the syntax of the picture contributes to the constitution of a representation of reality that is connected with the social environment within which the images are created. Narrative structures, as opposed to conceptual structures, contain vectors that are the visual proposition of the image syntax (Kress and Van Leeuwen 1996). For this reason, it is crucial to explore the narrative structure of this picture and to keep in mind its extremely important semantic aspect.

The action of the image is constituted by the vector, the line that connects all participants in one line. Refugees are moving as a mass and with a strong directionality. Usually, the actors of the image are the ones from which the vector generates. In this particular case, the refugees are the actor of the action and the action at the same time. Most importantly, there is no goal for the action. In the language of social semiotics this is a unidirectional non-transactional action, a process going in one specific direction that does not point at something in particular (either an actor or a goal). The absence of the goals is all the more relevant if its visual effect is combined with the strong directionality of the action given by the dark line of people walking through the territory. As Andersson has pointed out, talking about the visuality of migration securitization, "risk is made real through a world of arrows" (Andersson 2012, 9). The art theorist Rudolf Arnheim defines the represented participants in terms of "masses" or "volumes" with a" weight" and" gravitational pull". The vectors – that, as we have seen, represent processes – are also conceptualized as "tensions" and" dynamic forces" (Arnheim 1956, 1982). If we re-read the picture using the vocabulary of Arnheim, who was writing about visual perception, the meaning of a strong vectoriality without a goal may appear clearer. What the narrative structure of the picture is suggesting is that there is a mass and a tension in the direction of a familiar and quiet landscape.

---

[1] Salience in visual social semiotics indicates elements that are made more eye-catching than others. (Van Leeuwen and Jewitt 2001)

This vectoriality without a goal is one of the ways through which the depiction of refugees contributes to their being represented as threat: a salient arrow with no goals, and therefore no explicit objective of the action, launched into (our) peaceful and ordered territory.

This is one way to look at it. Since images are inherently polysemic for they obviously lend themselves to various readings and interpretations. For example, someone could be struck by a feeling of strangeness, given by the abnormality of such a walk, clearly not a Sunday stroll, a group of people moving through fields with children and rucksacks. At the same time, the strong vectoriality of the picture could also be read in terms of strength, as a symbol of people on the move, with strength and necessity to move. However, in a visual social semiotic perspective, the suggested meaning of an image is based on the combination of different layers of meaning which, combined, work to reinforce (when they underline different dimensions of the same narrative) or weaken (when they blend different or even antithetic accounts) a particular reading. In this sense, as discussed in a more detailed way in Chap. 3, reading a picture in a certain way is not just an arbitrary and personal interpretation. Rather, it is the consideration of the combination of the various levels of meaning of an image, seen in a precise geographical and historical context – in this case that of the European 'refugee crisis' and the mainstream media and political securitized account around it.

Vectoriality with no goal is not the only pictorial element contributing to the social imaginary of the refugees as threatening. The lack of eye contact with the represented participants makes this picture an 'offer' image. In this case, it offers information on what is happening in the Greek islands during the period of the so-called migration crisis. There is no establishment of direct contact between the viewer and the people portrayed in the image. The effect of distance is accentuated by two other photographic techniques (or semiotic resources in the language of visual social semiotics): the very long shot and the oblique angle. What all these elements do is to underly the social distance between the producer of the image (and the viewer) and the refugees. Distance between the audience – 'we' – and represented participants – 'they'– is kept at its maximum potential. Because of the importance of the *we* versus *they* narrative, a part of this chapter (see section *The Other*) is dedicated to the specific exploration of this dimension. What is important here is that the combination of different meanings included in the polysemic representation of the picture contributes to the representation of the refugees as a mass of people that, distinct and distant from us, has, nonetheless, a strong dynamic tension while we observe the group penetrating into our landscape.

Among the more than a thousand images collected for this study, this picture is not the only one displaying this specific narrative structure. The visual theme that I am calling *vectoriality with no goal* is present in at least other two dozen images, each of them in its distinctive way. While it is clearly not a dominant depiction of refugees in NGOs visual production, it is, nonetheless, present (especially in the visual production of CARE and MSF) and contributes to the reproduction of an account of refugees as threat.

There is little doubt that the intention of the image producer of these different, and yet sharing a "family resemblance" (Wittgenstein 1953) images had another goal in mind. Interviews with NGOs humanitarian and communication officers in the field showed that there are a variety of reasons behind their 'offer images'. In most cases the NGOs message focuses on information about the situation, their work on the ground and possibly the mobilization of emotional, civic, or monetary resources. In the case of the pictures above, the immediate accompanying text can be a useful indication of the kind of message that the organization had in mind. The short description complementing the picture briefly tells about the organization's work at the Greek-(North) Macedonian border and the situation that refugees are facing along the route (MSF 2015). The other pictures of this kind analyzed in the study were either accompanied by NGO statements against the closure of borders and protests against European migration policies[2] or part of photo-journalistic projects aimed at showing different dimensions of the migration paths (Save the Children 2016; Oxfam 2016b). However, the element of protest, or of the threats refugees are exposed to during their journey, are completely missing at the visual level from the pictures examined.

Of course, a single image cannot contain an infinite number of elements. This is an essential feature of the photographic genre: the framing choice influences what is present and what is absent.

In this arbitrary choice lies its power, a power made even more effective by photography's strong epistemological claim of representing reality (Kuhn 2013; Sontag 1973; Barthes 1981). What is critical in the discussion around vectoriality with no goal and its securitization potential is that in the all the similar pictures considered the element of protest, indignation, complexity of the situation are not visible. In a CARE picture in which the refugees in line are covered with heavy blankets which constitute the only visual hints about the 'uncomfortable' situation people are going through. At most, they function as an index of a cold, harsh climate, a lack of proper shelter, but the image adds little to the political and everyday conditions that Syrian refugees are faced by along their route and that is, instead, present and given relevance in the text accompanying the image. This is quite at odds with the widespread assumption that one image is worth 1000 words (Stocchetti and Kukkonen 2011). It is clear that one of the features for which NGOs use visual material is for its ontological "immediacy" (Hansen 2011). In talking about the relation between photography and the organizational mandate of bearing witness, the director of operational Support of MSF Athens Unit, affirmed: "We do speak out after bearing witness (not because we hear things in bars), but after we have seen things. If you can

---

[2] Text accompanying CARE picture: "#Refugees from #Syria and other places, women, children, and families, running from war-torn homelands, are entitled to safety and stability, not to be sent away from European borders."; MSF: "Thousands of men, women and children are stranded on the border between Greece and the Balkans after the sudden introduction in recent days of new control measures at the border on the western Balkan route: "The inability of European governments to find collective and humane solutions to this crisis only creates a chaotic, arbitrary and discriminatory situation." (author's translation from French).

accompany what you are saying with a pic is much more powerful and sometimes you do not even need to comment on a picture that speaks for itself."[3]

Despite the acknowledged importance given to the visual level from NGOs, there is a clear disconnection between the visual and the verbal representation of the long lines of Syrian people on the move. This disconnect is not something new in the account of situation of crisis. Campbell (2007) has shown how (textual) media accounts of the Darfur conflict have revolved around two dominant themes: genocide and humanitarian crisis. At the visual level, the political dynamics and implications of the Darfur genocide were either diluted in a simplistic and inaccurate representation of the systematic use of violence in an 'ancient' and 'ethnic' conflict (*ibid.*, 377), or completely obscured by the humanitarian crisis interpretation. One can recognize a similar dynamic in the *vectoriality with no goal* pictures. While the arguments of indignation, protest and need to act are strong and clear in the text, they are completely absent from the visual representation. As we have seen, the social distance put between the image producer and the people portrayed along with the intense directionality of the images' vectors pointing at nothing specific – if not pointing directly at the spectator – instead of underlying the humanitarian message, seems to undermine it. By only looking at the pictures, the representation of refugees resonates more with the media accounts of masses invading our space (Falk 2010), than with the reality of refugees stranded at borders on their way to escape conflict and its consequences.

5.2.2   The Rhetoric of Emergency (Fig. 5.2)

A mass of people, no faces, at night. It is the ultimate refugee emergency. This picture, published in October 2015, depicts a large group of people with no individual traits. In the lower left-hand corner there is a boy turning towards the viewer. Probably, he is looking at us, but his face is backlit, his gaze thus blocked. There is nothing going on, no action but only a mass of people on a dark night, lit only by a couple of lampposts. We can just guess that they are many. The weight of the mass is made heavy from the lighting from above and its position at the bottom of the image (Arnheim 1956). The lack of an action going on, or a vector pointing in some direction, reveals, by default, the "conceptual" structure of the image (Kress and Van Leeuwen 1996). Here the represented participants are portrayed "in terms of their more generalized and more or less stable and timeless essence, in terms of class, or structure or meaning' (*ibid.*, 79). More specifically, the mass of refugee is represented in its suggestive symbolic process. It eludes an analytical interpretation because of the impossibility of discerning individual features. Details are overlooked in favor of an "atmosphere" (Kress and Van Leeuwen 1996, 106) underlined by the extreme dark lighting and the blend of different colors in different tonalities

---

[3] Interview with MSF Director of Operational Support, April, 25th 2017, Athens.

Fig. 5.2 ©CARE, Facebook Syria Response October 2015

of brown. The de-emphasization of distinguishing characteristics and colors accentuates the symbolic value of the *carrier* – the mass of refugees – in its generic essence. The suggestive symbolic process of this image represents the participants as if their identity and meaning were originating from within them: a mass of people embodying the refugee emergency. The medium range shot, the back view and the lack of eye contact with any of the participants accentuate the social distance of the image producer and the viewer from the represented participants. The Facebook post accompanying the picture focuses on the difficult living condition of people on the move and the limited (open) routes available to them (CARE 2015). Once, again the information included at the textual level is not present at the visual level, where the theme of the emergency prevails.

While extremely recurrent in the media representation, the iconography of refugee emergency is not a predominant visual pattern in NGOs visual material, especially if we exclude boat pictures that constitute a specific visual theme (to which the following section is dedicated). More precisely, the iconography of emergency is a recurrent visual trope in NGOs representation of refugees in general and Syrian refugees in particular. But there are two different kinds of representation of the emergency that have distinctive visual patterns and are conceptually different. In a study on the manufacturing of the emergency by the Italian government following the arrival on the island of Lampedusa of around 30,000 Tunisians in 2011, Campesi (2011) identifies two different discursive regimes employed by the national authorities: the "securitarian emergency" on the one hand, and the "humanitarian emergency" on the other. While the former stresses the security aspect of refugee arrivals, the latter refers to the concept of human security. At the visual level, NGOs' representation of the Syrian refugee 'emergency' presents a similar dichotomy. In one case people on the move are depicted as threatening, in the other as referent objects of a threat.

Although clearly two sides of the same coin, I find both representations problematic but for different reasons and with distinct intrinsic dynamics. It is also important to notice that the largest portion of pictures portraying the situation of emergency are images of the Syrian war and its destruction. They therefore represent an idea of emergency based on the human security concept, not the one discussed in this chapter, but an emergency that is threatening people and causing their suffering and displacement (see Chap. 5 for analysis of NGOs representation of refugees as referent objects). Here, I am only focusing on the photos that frame the refugees' movements of 2015 and 2016 in terms of what Campesi has defined "securitarian emergency" and thereby contribute to their representation as threatening. These kinds of images, albeit not predominant, are at the same time not completely absent from aid organizations' visual narrative. While they are marginal in CARE, Oxfam and Save the Children aesthetic representations, they are slightly more common in that of MSF. This may have to do with MSF's emphasis on witnessing and speaking out to alert the public to abuses occurring beyond the headlines or to criticise the inadequacies of the aid system. However, also when intended to bolster or legitimise denunciation these images simultaneously contribute to reproduce an account of refugee movements in terms of urgency and emergency.

The concept of refugee emergency, so strongly connected with the so-called migration crisis between 2015 and 2016, is based on the idea of the exceptionality of the situation combined with the intensity of the phenomenon. The rhetoric of the refugee emergency as threat illustrates an extra-ordinary mass of people arriving in Europe. Particularly important in this perspective is the temporality of emergency that becomes a justifying argument for the state of exception (Agamben 2005). By reproducing this particular kind of emergency narrative, NGOs contribute to the constitution of a specific "social imaginary" (Calhoun 2010) of refugees that not only shapes our understanding of the phenomenon but also the action following that interpretation. The problem with this approach is that an emergency situation refers to what is going on without bringing into question its specific circumstances, causes and consequences (Calhoun 2010). This is problematic because "in contrast to the amorous relation, which is based on how something looks, understanding is based on how it functions. And functioning takes place in time, and must be explained in time" (Sontag 1973, 18). On the contrary, the rhetoric of the refugee emergency focuses on the here and now, concealing and preventing a more complex understanding, and reinforcing a securitization framework in which an extraordinary situation needs extraordinary measures that go beyond the ordinary political sphere (Buzan et al. 1998).

The emergency refers to something unexpected. The ontological unpredictability of the emergency takes everybody by surprise. Calhoun (2010) has shown how this widespread feeling of unpredictability is especially reinforced by the media. Whereas more complex analysis may appear in print, news, and TV news in particular, represent emergencies as springing from a vacuum. On this point, it is quite clear that the logic of NGOs visual representation works in a different way. As we have noticed before, photos of the level of violence and devastation in Syria and the difficult conditions of refugees in Jordan, Lebanon, Greece and further north along the Balkan route constitute the large bulk of NGOs visual production. However, although NGOs visual representation does not conceal the causes producing the crisis, it generally fails to connect them with the resulting 'emergency' at the visual level. Aid organizations' visual material is compartmentalized as is its practical functioning with different field operations responding to different 'emergencies': whether it is winter in Lebanon, perilous landings on Greek shores or being stranded in Serbia. As a humanitarian worker I interviewed in Greece affirmed: "here in Greece the mission of Save the Children is not too big. We are only responding to the refugee crisis. There are no projects on integration, or more development programmes like in other countries. When the crisis will be over, we will be probably pulling out…or at least I believe so. There was no Save the Children Office in Greece before".[4]

At the visual level, this aspect is particularly evident if we look at the compositional meaning. We may take Save the children's website at the time of the refugee 'crisis' as an example, although similar considerations have emerged from analysis of the websites of all the other three organizations. First of all, there is no link to a page devoted to the war in Syria and its consequences for Syrian people in the country and abroad. The user is left to navigate the site either by topic in the section 'what we do' (in this case a list of the NGOs' sectors of intervention), by news, or by country in the 'where we work' drop down menu. Clicking on the 'read more' button on the homepage opens an article on children besieged in Syria. There are no links or visual connection with the refugees in Europe or in neighboring countries. Navigation through the Syria-dedicated page offers a similar pattern. There is no link, literally and conceptually speaking, between what is going on in Syria and the refugees' displacement.

In the opposite way, but with the same result, the photographic essay devoted to Children on the Move in Europe first published in September 2015 and last updated in July 2016 (Save the Children 2015) does connect what is happening in Syria with the situation of displacement, mentioning, through photos and text, conditions in Syria, Lebanon, Jordan, Serbia and elsewhere. Yet, this essay includes refugees from Afghanistan, Egypt and Somalia, putting all children on the move in the same frying pan, as if their specific contexts and reason for displacement did not matter. The result is somehow similar to what Nyers (1999) has noted regarding the United Nations High Commissioner for Refugees' (UNHCR) website. The page devoted to showing *What it is like to be a refugee?* gathers pictures of different displacement situations, including Bosnia, Kosovo, Rwanda, Tajikistan and Vietnam. The different photos portray the distinct dynamics and complexities of each of these refugee movements. Yet, they are gathered together because of the shared experience of

---

[4] Interview with a Save the Children Field Manager, 24th March 2017, Athens.

refugeeness. This "universalist, humanitarian perspective"' (Nyers 1999, 16), visually symbolized by the cover image of an empty shirt hanging to dry outside an emergency shelter, suggests that all different displaced people are unified by the same feeling of lack, of emptiness instilled by the empty shirt hanging over without a body inside. Such visual compartmentalization creates a disconnection between the causes of suffering and displacement and their consequences. They are represented as suddenly and surprisingly happening, or, on the contrary, the conceptualization of refugeeness in its depoliticized and timeless emergency essence.

All this is completely ad odds with the comprehensive knowledge of the Syrian situation that NGOs should have presented, given their large operations in Syria's neighbors (all four NGOs also work inside Syria) and in Europe for at least the previous two years. This compartmentalization is also in conflict with the role that NGOs have in the management of complex emergencies (Dillon and Reid 2000; Duffield 1994). UNHCR defines complex emergency as large scale events, intensified by armed conflict (internal or international) causing serious human rights violations and large scale suffering for the civilian population, producing large scale displacements (Ogata 1993). Most NGOs interpret their mandate to address 'root causes' and protect rights in a holistic manner typical of the rights-based approach. However, at the visual level, this is not represented at all. The response is compartmentalized, and the different situations result in disconnected reproducing of the 'social imaginary' of emergency as unpredictable and disjointed from its complex net of causes. Such emergency representation has important implications when it comes to its relationship with human rights and their individual/collective dimension. The depiction of displaced people in terms of invisibility and acorporeality erases any individual trait and makes it difficult to think in terms of political and social rights for a collective and indistinct group (Nyers 1999).

Another aspect of the concept of emergency, and the refugee emergency in particular, is its representation as an exception to a given order, an assumption that the disorder is caused by local factors (Dillon and Reid 2000). In the forward to *The State of The World's Refugees 1993: The Challenge of Protection,* Sadako Ogata, the former UN High Commissioner for Refugees, affirms: "The subject of refugees and displaced people is high on the list of international concerns today not only because of its humanitarian significance, but also because of its impact on peace, security and stability. The world cannot reach a new order without effectively addressing the problem of human displacement" (Ogata 1993, 2).

There are numerous different understandings of 'order' in which global displacement represents disorder and exception: a matter of citizenship as the "authentic ethico-political identity" (Nyers 1999, 3), that which eludes "a predictable system of relations and flows' (Calhoun 2004). While order is associated with development, disorder is associated with underdevelopment (Duffield 2001). Images of masses of people visually represent this *exception* to the normal flow of things and ordinary people's movement. Against this background, NGOs become key actors of containment and management of such disorder. At the visual level, this representation is reinforced and reproduced by the disorder of the masses or the need to 'manage' them. What these pictures fail to represent is that refugees, far from being the exception to an allegedly stable international order, are, rather, part and parcel of a "world-systemic phenomenon" and even more, of the "national order of things" (Malkki 1995). As Malkki has pointed out, this essentialized representation of refugees as a mass of displaced person sharing the refugee experience as a group reinforces the vision of the nation state as the "natural or necessary order of things" (*ibid.*, 511). She argued that this mutually constitutive relation between a functionalist and essentialist representation of refugees and the assumption that state sovereignty is *the* order have two important consequences. It not only naturalizes closed border policies but also the need to 'manage' refugees who are 'out of place'. The idea of emergency and disorder is threatening because it produces instability in an otherwise supposedly stable situation. In this representation of the emergency, refugees are the disorder, refugees are the threat.

Of course, NGO framing of Syrian refugees' arrivals in Europe in terms of emergency can also be looked at within a logic of economy of attention. Many commentators have explained how NGOs in the contemporary news production landscape compete within this economy (Friese 2017; Fehrenbach and Rodogno 2015; DeChaine 2002; Cottle and Nolan 2007; Nolan and Mikami 2013; Dijkzeul and Moke 2005). However, although this representation may prove effective for drawing public attention, raising awareness, and mobilizing resources for solidarity interventions, it remains intrinsically problematic. In competing for attention, it reveals the exceptionality of the situation, the need of emergency measures (Friese 2017), thereby reinforcing and reproducing a securitized image of people's movements.

At the theoretical level, the compartmentalization of several different emergencies, very much in line with the structures through which NGOs operate in the field, and the representation of refugees as anomalies to be managed, seems to perfectly fit within the Copenhagen School's conceptualization of securitization. Its framework is, indeed, based on the theoretical assumption of the exceptionality of security politics (Huysmans 1998) – in this case emergency interventions for situations that escape ordinary political management. In including humanitarian agencies in the realm of potential securitizing actors, Watson has also suggested seeing securitization as a continuum on "a spectrum of exceptional/institutionalized" (Watson 2011, 8). Such approaches seem to miss something crucial about securitization and the production and reproduction of emergencies. As the Paris School – a postmodernist approach to security studies inspired by Michel Foucaut and Pierre Bourdieu – and Didier Bigo in particular, have perceptively shown, this perspective fails to address the issues linked with the "effects of power that are continuous rather than exceptional"' (Bigo 2002, 73). Although the Copenhagen School obviously does not agree with the assimilation of migration with security, as Bigo notices, they "accept the 'truth' about what security is not in the way they agree with the military (Waever in particular is critical of the existential character of the threat), but do so by accepting the framing of a different domain of security beyond the political—one linked with emergency and exception (Bigo 2002, 73). The way NGOs visually represent emergency works with a similar logic, where the narrative of emergency fuels a securitization discourse. Both are presented in terms of exceptionality.

Overall, the representation of the arrivals of Syrian refugees in Europe as an emergency has several negative implications. It overestimates the impact of Syrians seeking refuge in the West while at the same time neglecting the magnitude of displacement within Syria and in the Middle East. It overlooks the causes of Syrian people's movements from their homeland and onwards from neighboring countries. It leaves politics and economy out of the equation. It neglects the role that Europe and the USA has played (or avoided playing) in the unfolding tragedy of the Syrian uprising. It paves the way for emergency solutions that are by definition short-term and aimed at the immediate management of the disorder. I find the presence of pictures of this kind – even though residual – surprising for two main reasons. The first is linked with the intention, expressed by a few of my informants, that one of objective of their NGO's communication was to challenge the stereotypes around the refugee issue by showing the humanity of displaced people. Pictures like the one discussed above seem to go in the exact opposite direction, since the visual rhetoric of emergency resonates more with sensationalist media headlines about refugee 'emergency' and 'crisis' dimension of refugees' movement, rather than its individual and human experience. The second reason is related to the fact that all images visually representing the refugee emergency were depicting refugees in Europe. No such images were used to represent Syrians crossing borders into Jordan, Lebanon or Turkey, thus obscuring the reality that in September 2016 they were hosting nearly five times the number of Syrians as those who had reached Europe (Migration Policy Centre 2016). During interviews, informants often underlined how Lebanon, Jordan and Turkey were bearing the brunt of the burden of Syria's exodus[5] and how European countries needed to do their fair share in terms of asylum[6] and resettlement.[7] However, this important message has been visually completely overlooked. The result of the representation of Syrian arrivals in Europe in terms of emergency has the effect of securitizing the refugee issue. The problem is that this representation "is not merely a description of the world, more or less accurate, but an abstraction that plays an active role in constituting reality itself" (C. Calhoun 2004, 391).

5.2.3 *Boats, Refugees at Sea, and Rubber Dinghies (Fig. 5.3)*

In the social imaginary of emergency, the iconography of boat and so-called boat people[8] is particularly recurrent and relevant. Pictures of people on rubber dinghies became quite common between 2015 and 2016, illustrative of the journey undertaken

---

[5] Interview with CARE Regional Communications Officer for the Syria Response, 23rd May 2017, Amman.

[6] Interview with Oxfam, Policy Advisor, 8th March 2017, Beirut.

[7] Interview with Save the Children Information, Communication and Media Manager, 24th April, Athens.

[8] Drawing on Pugh's (2004) use of the term "boat people"' I use this concept to identify their status at the legal and discursive level. I do not imply any generalization on their specific origin and causes of movement. The term was popularised to denote people escaping South Vietnam following the US defeat in 1975 (Kushner and Knox 1999).

Fig. 5.3 ©Save the children/Anna Pantelia

by many people on the move, and specifically Syrian refugees. At the end of 2016 Syrians still constituted the greatest proportion of those landing on European shores, one fourth of the total, but the number drastically decreased from 524,597 people between January 2015 and January 2016 to 81,949 at the end of 2016 (UNHCR 2016). The drop in the numbers is largely explained by the coming into force of the EU-Turkey Agreement signed in March 2016 which, as noted, effectively closed the Balkan route into Europe.

A picture published by Oxfam on Facebook in 2016 (Oxfam 2016c) shows a small rubber dinghy overcrowded with people and two humanitarians recognizable by their high-vis jackets. Although we can identify heads, refugees are fused together in an intricate bundle of bodies, while their 'saviors' stand out. As with pictures examined above, refugees are presented as an undistinguishable group. We can neither see their faces nor facial expressions. Some orange life jackets stand out as symbols of the perilous journey these refugees have had to undertake on the sea leg of the Balkan route. A few black lifebuoys float in the background, symbol of rescue. The setting of the image helps us to situate the scene in time and space. It is twilight in what is for Europeans a familiar Mediterranean seascape. There is an action going on: the vessel is arriving, its passengers poised to disembark. The vector points toward the shoulders of the viewer. At first it may seem that the boat tends toward the rescuer, but closer inspection shows that he is moving toward the side of the boat, probably where his colleague is standing. Delving deeper, we can spot a rope on which the person in front of the dinghy is leaning, in the act of pulling. The visual non-transactive action can be translated as 'refugees arriving on (our) coasts. The visual emphasis on the humanitarian staff draws attention to an embedded process of rescue. Although the two people with high-vis jackets are not doing anything specific, it is quite clear that they are part of the solidarity effort, either volunteers or professional, engaged in providing 'emergency' care, NGO bodies on the ground. The lack of eye contact and the frame size connote this photo as an offer image.

While providing us with the visual information about what it is going on, the medium range shot puts a certain distance between the represented participants and the viewer (us). However, this sense of distance is contrasted by the frontal angle and the directionality of the vector that keep us engaged, at least to a certain extent, to the image. Along the "close personal distance" – "far social distance" continuum identified by Kress and van Leeuwen (1996) this image situates us at a relatively close social distance. What the combination of these semiotic resources suggests is that groups of refugees are arriving toward us and that this is something that implicates us in one way or the other, whether we like it or not.

The words 'THIS IS NOT ABOUT REFUGEES. THIS IS ABOUT PEOPLE.' superimposed in caps on the picture is given explicit saliency. The meaning of the caption (and the intended message of its image producer) seems clear: there may be questions about the legality of their movement, but this is a question of humanity. The caption strikes me as problematic for various reasons. Firstly, it is linked with adversative construction of the period. Why is the notion of people opposed to that of refugees? What is inherently good about the former group that is, on the contrary, bad about the latter? One would guess that the message's intention is to desecuritize the refugee issue, intended following Huysmans (1995), to be a narrative that tries to convince that migrants are not a security threat. An Oxfam policy advisor told me that their "focus is on human rights and people issues. We do not talk about security issues at all. On the contrary, our job is to remove the security argument form the equation'.[9] Although she was not referring specifically to the representation of 'boat's people', this was the answer to my question on how the organization was dealing with questions of security around the refugee issue and its accommodation with the humanitarian message in its advocacy strategy. Despite the stress on the idea of shared humanity – the rhetoric of the 'we are all human after all' can be interpreted as a sincere 'objective desecuritization' effort which plays on victimhood and solidarity. Its meaning remains, I believe, quite problematic. By trying to convince us that refugees are not a security threat, this narrative reproduces the native/refugee dichotomy and in this way (re)securitizes the refugees. But what it is even more important is that the text appears at striking odds with the ideological value of the image. The indistinctiveness of the group of refugees on the dinghy works exactly in the opposite direction, taking away individuality and humanity from the people represented.

The picture above is clearly not the only one of its kind. Between 2015 and 2016, images of boats have been particularly present in MSF visual material and, to a much lesser extent, in Oxfam and Save the Children. CARE seems not to have published a single image of migrants' boats during the observed period. It is not a surprising finding that this visual theme is more present in MSF communication, since the NGOs has explicitly decided to capitalize on media and public attention on maritime SAR operations to draw public attention to the Mediterranean border regime (Cuttitta 2017). The absence of images of boats in CARE's visual material,

---

[9] Interview with an Oxfam policy advisor, 8th March 2017, Beirut.

and the limited number in Save the Children photos, seems to be mostly linked to their attempt to avoid direct implication in highly politicized debates. This would explain a similar attitude despite the two NGOs' respective absence (CARE) and presence (Save the Children) in the waters of the Mediterranean. With the same logic, but opposite result, the presence of such kinds of pictures in Oxfam's communication – even though the NGO has not worked in SAR – can be explained by the agency's attitude of considering political confrontation an option when the topic is linked to human rights.

In recent decades the iconography of overcrowded boats has become a symbol of large migration flows and a visual trope quite familiar to Western audiences. They have included Vietnamese adrift in the South China Sea after the victory of the Vietcong in 1975,, Albanians disembarking in Brindisi in 1991, boatloads of asylum seekers trying to reach Australia, Cubans crossing to Florida, Somalis in the Gulf of Aden trying to reach Saudi Arabia Africans crossing the Straits of Gibraltar en route to Spain and, most recently, refugees at peril crossing the English Channel. . Images of overloaded boats become symbolic to such a point that a famous Italian fashion photographer used one of these images for a commercial campaign.[10] In the media (Falk 2010; Friese 2017; Hermanin 2017) and in political accounts (Andersson 2012; Gale 2004) images of cramped vessels very often serve as visual background for discourses on irregular migrations, and invasion.

In order to better grasp the meaning of such widespread use of images of overcrowded boats in the public debate on people on the move, it is important to consider the strong ideological value attached to this iconology. Vessels full of people inspire feelings of threat and fear (Pugh 2004; Falk 2010; Bleiker et al. 2013; Musarò 2017; Friese 2017; Furedi 2005; Massumi 1993). In a content analysis of Australian media representation of asylum seekers, Bleiker et al. (2013) have observed how the large majority of pictures privileged medium/far range shots where refugees were represented as indistinct groups and how the recurrent presence of a (distant) boat in many images reinforced this visual pattern of emotional distance. The semiotic resources at play in those pictures, exactly as in our case, were framing the issue in terms of threat, thus fueling security discourses.

The consequence of this kind of depiction is that it dehumanizes the represented participants. This point is particularly significant if we go back to the Oxfam dinghy image and its caption. The scope of the verbal message was clearly intended to desecuritize refugee arrivals and bring the issue back to a more human (or also, in this case, humanitarian) ground. However, the need to rescue people at risk of drawing in the Mediterranean and the intended solidarity message end up being diluted. Indeed, the visual and text semiotic signs of the image steer away from this aspirational narrative to actually reinforce and reproduce a dehumanized representation of refugees. The social distance and the dehumanization contribute in fact to a framing of Syrians arriving on European coasts in terms of threat.

---

[10] Oliviero Toscani, 1992 picture of an overcrowded boat for a Benetton campaign.

On top of the meaningful iconology of the boat, the idea of the ship is linked to its natural environment: the sea. The texture of water, with its fluid, infiltrative power is present in several liquid metaphors on migration. People on the move have been commonly described in terms of floods, waves, and flows (Pugh 2004). This representation, both verbal and visual, resonates not only with the emergency frame discussed above, but also with discourses of infiltration, penetration, and invasion, one of the founding elements of the depiction of refugees as threat.

The interaction between 'boat people' and NGOs in the context of the "cosmopolitan space" (Pugh 2004, 51) of the sea, was particularly relevant during the 2015–2016 refugee 'crisis'. NGOs have been repeatedly accused, particularly by populist politicians in Italy and Spain, of facilitating the work of human smugglers and human traffickers operating in the Mediterranean (Huffington Post 2016; Sea Watch 2016; Financial Times 2016; Liempt 2016). While emergency organizations such as MSF and SCF were engaged in maritime SAR operations to reduce the number of people drowning – 6281 in 2015 and 7932 in 2016 according to the International Organization for Migration (IOM 2017) – they have been the subject of a concerted campaign to criminalize them.

Implicit in the accusation was the assumption that the actions of NGOs were somehow synonymous with those of people smugglers, both facilitating the 'irregular' crossing of international borders.

The conception of people on the move as a threat was certainly not part of the narrative of aid organizations as, on the contrary, they sought to rescue those at peril on the sea. However, visually, the iconography of boat people and its strong ideological value has not been particularly challenged by NGO-disseminated imagery consistently presenting an ensemble of semiotic resources very similar to those used by media in the photographic repertoire of 'boat people'.

At the same time, NGOs representation of people at sea and boats offers visual elements that are not usually present in media accounts. While reproducing and reinforcing a securitized representation of refugees as threat, NGOs also introduce features that slightly differ from the dominant iconology. One innovative element is the presence of 'rescue people' among the represented participants. Although medical and security personnel are often portrayed in pictured of disembarkation (see pictures in Friese 2017; and Falk 2010), this mostly happens in close range shots or portraits of 'victims' rescued by 'heroes'. In this case instead, rescue staff are included in medium-far range shots where it is nigh impossible to recognize individual traits. Whether humanitarian operators are recognizable or not, they are usually part of the image's setting and included as an information element symbolizing the valiant presence of NGOs on the humanitarian front line. They are often accorded prime attention via the saliency of their highly visible yellow jackets or, in the case of the picture above, by being foregrounded. Interestingly, their position between the refugees and the viewer, positions the role of the organization between the people on the move and the spectator and to a certain extent mediates their relationship, inspiring an emotional response completely different from the one evoked by the direct eye contact typical of other kinds of visual themes. However, this does not alter the dehumanization of the refugees on the boat. As Butler has shown, "the 'frames' that work to differentiate the lives we can apprehend from those we cannot (or that produce lives across a continuum of life) not only organize visual experi- ence but also generate specific ontologies of the subject. Subjects are constituted through norms which, in their reiteration, produce and shift the terms through which subjects are recognized. These normative conditions for the production of the sub- ject produce an historically contingent ontology, such that our very capacity to dis- cern and name the 'being' of the subject is dependent on norms that facilitate that recognition" (Butler 2009, 3–4).

Something similar happens in a Save the Children picture. The classic iconology of the boat crammed with people is challenged by two elements: the subject in the foreground and the point of view. A baby girl in the arms of a woman, possibly her mother, perhaps a volunteer present at the disembarkation, partially covers the rest of the view. This element attracts our attention because of their centrality in the middle of the frame, the close-range shot – atypical for this visual theme – and the pink color of the baby's coat in stark contrast with the usually more somber colors of boat pictures. Although this image introduces new elements like childhood, care, and the need for protection (the arm hugging the baby) those portrayed remain unidentifiable. There are no distinguishing traits, no eye contact, no indication in the text of who the baby girl, the woman and the people on the boat in the background are. They are refugees, dehumanized in their timeless essence. As Bleiker, drawing on Malkki, has argued, "we see no faces, no real people. We see just anonymous masses. We see an abstract and dehumanized political problem. Such pictures sup- press or overlook the types of factors that make people human" (Bleiker et al. 2013, 411).

Another innovative element of this particular image is given by its interactive meaning. Instead of the lateral or aerial angle, typical of the boat iconology, this pic has been shot with a high angle and a frontal perspective. The boat comes towards us, without interruption, practically on us, strengthening the meanings associated with the vectoriality with no goal discussed above. The high angle puts the viewer in an "imagined position of power" (Kress and Van Leeuwen 1996) that does not contrast with the interactive distance inspired by the other semiotic resources at play.

It is also important to note that the different articulations of the visual theme of the boat and boat people produced by NGOs introduce another component, one that we could not see in the classic media images: the harbor, the place of refuge, the maternal embrace, the rescuers. What all these different visual clues lead us to think is the presence (or need) of some level of protection. This elusive, but still clearly visible, element is probably more in line with the verbal narrative of aid organiza- tions that work to assist people at sea because they deemed *at* risk, and not because they constitute *the* risk.

It is undeniable that boat and emergency pictures alike contain elements evoking, at least partially, the difficult experience of forced displacement. This is clearly one of the intentions of image producers when it comes to NGO's communication.

Showing what is happening, appealing to feelings of shared humanity, challenging stereotypes around refugees, exposing their being at risk, to cite just a few. Referring specifically to the pics of boats and boat people an NGO communication manager told me that the reason they were deployed was because the problem was still ongoing, they needed to inform people what was happening and what the organization was doing as part of SAR activities.[11] Many NGO communication and advocacy officers I spoke to indicated that these reasons were fundamental to their dissemination strategy. Of this there can be no doubt. This is not in question.

Nonetheless, aid organizations' visual representations of boat people and the refugee 'emergency' between 2015 and 2016 remain, I believe, problematic. Some humanitarian practitioners acknowledged this when talking about their communications material in general terms. As one communication officer told me: "what is happening in Greece is very bad because NGOs have a lot of power in producing the discourse around refugees, but they are doing a very bad job (…) For example there is an NGO, I will not tell you the name, that recently did an exhibition and it was full of people on boats, very bad conditions. Of course, it could be useful to show the condition of refugees' arrival but not now. It is not 2015 anymore.[12] Even though she acknowledged the importance of the witnessing role of NGOs and the need to inform the public, she also recognized that this depiction was also contributing to a distorted representation, at least at the moment of our conversation in June 2017 when large-scale arrivals of people on Greek shores were practically over.

However, when the NGOs intention to inform people about the situation is taken into account, the representation continues to be ambiguous. As Cuttitta (2017) has pointed out, the visuality of boats and SAR operations plays an important role in the way NGOs intend to disseminate a 'correct' image of migration. These images are important for documenting events and showing donors how their money is being used. Efforts to challenge securitized representations privileging closed-up representation of women and children in need is also evident. However, to some extent, these pictures "end up perpetuating, the neo-colonial image of the 'good' Europeans helping the suffering victims of the 'bad' smugglers. Thus, they contribute to portraying migrants as individuals in need of help in the first place, rather than as subjects who, in trying to realize their projects, are contesting and defying the political construct of the EU border regime" (Cuttitta 2017, 13).

In short, the images of boat people produced by NGOs during the so-called refugee crisis are not subverting the visual description of refugees arriving on *our* shores as threatening. The blurred blend of bodies, the indistinctiveness of the represented participants, the 'vectoriality with no goal', the frame size, perspective, and point of view, all work together to foster inquietude and fear, reproducing and reinforcing a securitizing discourse that depicts refugees in terms of threat.

---

[11] Interview with MSF Communications Manager,25th April 2017, Athens.

[12] Interview with CARE Emergency Communication Officer, 13th June 2017, Athens.

5.2.4   Conceptual Structures (Fig. 5.4)

Throughout the visual analysis of NGOs images, there has been something that kept preoccupying me with regard to aid agencies' representation of Syrian refugees. It was a feeling inspired by very different pictures, including a portrait of single individuals as much as of groups of refugees. It was not about the subject *per se*, nor about a particular visual theme or a specific setting. Then I realized: it was something about the use of conceptual representations (Kress and Van Leeuwen 1996). These kinds of visual structures differ from the narrative ones that are constructed around an action or a process in the making. Conceptual representations formulate the participants in their abstract term, either "in terms of class, structure or meaning" (*ibid.*, 79). I have already discussed this kind of depiction in reference to the mass of refugees often photographed as symbolizing the refugee 'emergency'. Yet, the significance of the conceptual structure and its use instead of a narrative one can tell us much more. The choice among the two patterns "is important, since the decision to represent something in a narrative or conceptual way provides a key to understanding the discourses which mediate their representation" (Jewitt and Oyama 2001, 141).

The four organizations rely on this kind of representation to a different extent. CARE uses conceptual structures more than any other NGO. Save the Children and Oxfam also use them extensively but tend to always include a fictitious name and

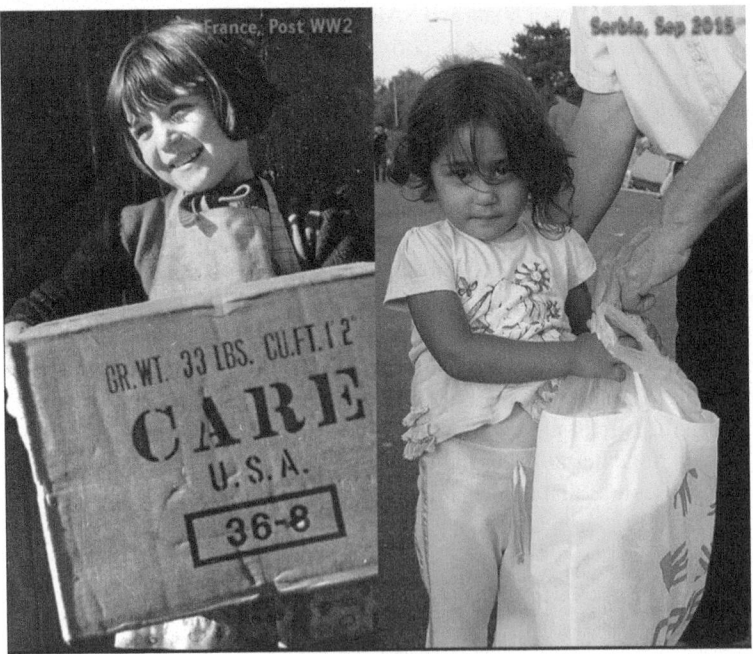

Fig. 5.4 ©CARE, Facebook Syria Response 29 September 2015

age (with no other substantial details) in the caption in an attempt to enhance the feeling of personal connection. MSF is the organization using these kinds of images to the least extent.

A closer look at the implications of the use of this specific semiotic resource in many and extremely different pictures helps us to understand another way through which NGOs contribute to a representation of refugees as threatening.

CARE's picture is a very good example of conceptional structure for it combines at least two different kind of conceptual processes. The image is composed by the juxtaposition of two photos. On the left, a black-and-white portrait of a child, a girl, smiling while holding a large box, the provider of which is clearly indicated by the prominent writing at the forefront: CARE, U.S.A. We can clearly induce that the box contains useful assistance items. We also read that the box must be quite heavy, some 15 kilograms, a challenge for such a young child to hold. But this is a detail, one which we may assume was a detail also for the image producer. The girl holding the box exemplifies the victims of war receiving assistance from the USA in post-World War II France. On the right, alongside it we have a color portrait of a girl, probably slightly younger, looking at us and smiling while holding a heavy bag whose origin is also quite obvious because of the CARE logo. Again, the bag is heavy, testimony to CARE's largesse and generosity. It is so heavy that an adult needs to help hold it while the girl poses. Text in the top right corner locates the image in Serbia, in September 2015, during the peak of the refugee 'crisis' as Syrian refugees were moving along the Balkan route. The two images differ in tiny details that mostly have to do with the changes of humanitarian (and technical) communi-cation styles over the years: black-and-white versus color, the direction of the girls' gazes, the apparently accentuated realism of the contemporary image. The differ-ences do not go much further, obscuring the reality that the situation of civilians in post-war France and of displaced Syrians on the move are strikingly different.

What connects the two images is their depiction in terms of a timeless essence of victimhood and humanitarian charity. Two visual social semiotic processes at play here reinforce the conceptual structure of this representation and support the above-mentioned interpretation.

The first one is a 'classification process', and more precisely a quite paradig-matic example of what in visual semiotics is known as a 'covert taxonomy'. In this kind of structure, the represented participants are related "in terms of 'a kind of' relation'" (Kress and Van Leeuwen 1996,79). Their juxtaposition stands for their belonging to an encompassing category. This particular image is a covert taxonomy as the overarching typology is not explicitly visually indicated, although is revealed in the accompanying text: "#CARE has been *delivering hope* in the form of care packages for 70 years now! *From victims of World War 2 to victims of the #Syria crisis*, CARE cares" (emphasis added). Not surprisingly, Kress and Van Leeuwen's description of covert taxonomies seem to exactly describe our image: "One visual characteristic is crucial in the realization of covert taxonomies: the proposed equiv-alence between the Subordinates is visually realized by a symmetrical composition. The subordinates are placed at equal distance from each other, given the same size and the same orientation towards the horizontal and vertical axes. To realize the stable, timeless nature of the classification, the participants are often shown in a more or less objective, decontextualized way. The background is plain and neutral. Depth is reduced or absent. The angle is frontal and objective. And frequently there are words inside the picture space." (Kress and Van Leeuwen 1996, 79). Moreover, the classification obviously implies an ideological meaning. The classification does not mirror a natural or real relationship. Instead, "it is the structure of the image that naturalizes it. And "naturalization" is not natural, whether in images or in language. The ordering in the image itself produces the relations'.(*ibid.*, 79).

The second process at play is a symbolic attributive process. Such kinds of structures typically include a human subject depicted for what s/he (symbolically) signifies – the carrier – and whose identity is defined in relation with the second element – the symbolic attribute – that stands for the meaning. Symbolic attributes are easily recognizable by specific characteristics identified by art historians such as their saliency, their looking in some way out of place or being typically associated with symbolic value (Kress and Van Leeuwen 1996). In this picture the girls clearly stand for *all* victims everywhere, as also indicated in the text. The donation boxes represent the humanitarian assistance, their 'hope' in the rhetoric of the text accompanying the photo.

There are other examples of how conceptual structures can be used in very different visual themes: a medium range shot of family in a shelter, the close -up portrait of a veiled woman, an image taken from behind of a little girl walking in a refugee camp, and the portrait of a man with his daughter/niece encaptioned 'Syria Refugee crisis'. Undoubtably, these images differ from other kind of representations of refugees as threats: represented participants' faces are mostly recognizable and they look directly at the camera (and therefore at the viewer). However, what all these pictures have in common is the essentialization of the refugee experience into the image. The people represented, although clearly visible and identifiable, are not the subject of the pictures. They are symbols of Syrian displacement.

Although such images are extremely different from the ones typical of the most recurrent threatening visual themes (boats, masses of refugees and vectoriality without goal), the semiotic resources at play here equally contribute to the dehumanization, de-contextualization, and de-politicization of the refugees. The abstractness of the refugee masses here works in a similar way as the abstractness of the individual portrayed. Of course, this shift from the represented participants described as a group of unidentifiable people to the depiction of one individual, a parent with the child or a family, is important and should be acknowledged. It is probably intended to humanize the refuges, giving them a face, a voice, something that we (the viewer) can emotionally relate to. These pictures, extremely common in contemporary humanitarian visual representation of refugees, are intended to bring back the humanity that was lacking in Malkki's (1996) stereotypical images of a sea of humanity. However, at the same time, the individualities, names, histories, experiences and feelings of the people portrayed are taken away by the photographic selection of a conceptual structure. They stand there as abstract symbols of the overarching category of Syrian refugees. But that it is not all: they also, not surprisingly, represent the exact categories of vulnerabilities of humanitarian work: women and children, families, older people, people with disabilities.

These kinds of images contribute to a representation of the refugees as threatening for two main reasons. On one hand, conceptual structures facilitate an association between the refugee issue and the people depicted to represent it. Noticeably, such people are children, especially girls, women, older people and families. It seems that they can be the poster images of refugeeness precisely because they are not usually threatening characters. On the contrary, they usually are victims, the threatened ones. This is probably one of the reasons why they are chosen by NGOs to embody the entire category: chosen for their being unthreatening. Yet, by coming to symbolize the entire refugee population, they visually completely exclude the adult males who account for at least half of the Syrian displaced.[13]

These representations, therefore, reproduce and reinforce a conceptual dichotomy between the victims, the weak, the threatened and the others. They are the 'good' faces, because they are not a threat. But what about the others, the majority of the refugee population? It is not the case that adult Syrian men do not appear in conceptual structure images. This a key point. On the other hand, these pictures become the symbols of the refugees, when we want to talk about the people and not the emergency. The photos have the effect of visually creating a refugee typology that as much as it intends to establish emotional connection can produce the 'other'. Above all, this is true when the symbolic carriers are represented as aliens, especially when one pays attention to the symbolic attributes present in the images: the woman' hijab, the man's headgear, the dirty and muddy environment through which the girl is wandering. All the images using a conceptual structure produce refugee poster images as a category of people different from us. The next section will explore this dimension, discussing the production of self-identity and the other.

5.2.5   The Other

A Oxfam picture of Moria refugee camp published on Oxfam website in 2016 (Oxfam 2016a)portrays a quite large group of people sleeping or resting on the ground in an open field. Within the mass of people, we can distinguish few faces: in particular a man sleeping in the foreground with his hand on his head, a woman looking aslant at us. We can distinguish few heads as people blend seamlessly with their baggage, sleeping bags and blankets. We have to look carefully to see where one body ends, and another begins. It is a mass of people on the move – but temporarily immobile – and the mass completely fills the frame. The woman's gaze is at the center of the narrative structure. She is center stage, and salience attributed to her figure by the saturation of her hijab's color and her position at the top of the

---

[13] For detailed statistics refer to http://data.unhcr.org/syrianrefugees/regional.php [last accessed on the 12th January 2017] for data on the Middle East and https://data2.unhcr.org/en/situations/mediterranean/location/5179 [last accessed on the 12th January 2017] for data on the Mediterranean.

prospective triangle formed by the two men' knees at her side. Despite the presence of eye contact, this is not the ideal type social semiotics considers a 'demand image'. The perspective, the angle and the distance of the shot, all suggest that this picture is rather an information image. In fact, it offers insights into the situation of refugees recently arrived at Moria. The picture tells us that there are many people, left on the ground in a cold climate and probably waiting for something. The accompanying text completes the information: "In March, as a result of a large increase in people arriving in Moria, the Greek authorities transformed the reception facilities into detention centers. People are now being held pending their mass return to Turkey, following the deal struck between the EU and Turkey. Furthermore, the Macedonian border crossing, where thousands of people had been arriving from Moria Camp to obtain authorization to cross on foot into the former Yugoslav Republic, has been closed. There are now more than 50, 000 refugees currently stranded in Greece" (Oxfam 2016a).

What this picture also does, and it is particularly relevant for the argument here presented, is that it creates enough distance between the represented participants and the viewer to mark a neat line between the refugees, the 'other', and 'us', the viewer. To a Western audience the hijab immediately evokes something alien, but it is not only this detail marking a distinction between 'we' and 'them'. In the production of an image, the choices regarding the perspective, the angle and the frame size imply the possibility of expressing subjective attitudes towards represented participants, human or otherwise" (Kress and Van Leeuwen 1996, 129). Most importantly, "By saying 'subjective attitudes', we do not mean that these attitudes are always individual and unique. We will see that they are often socially determined attitudes. But they are always encoded as though they were subjective, individual and unique" (*ibid.*, 129). The picture is taken from an oblique back view that encodes detachment (as opposed to the involvement of the frontal view). Kress and van Leeuwen have pointed out that "the oblique angle says, 'What you see here is not part of our world; it is their world, something we are not involved with.' (*ibid*, 136). In this case, the head of the woman coincides with the perspective's vanishing points. Her gaze could contrast the feeling of estrangement suggested by the perspective. However, as we have noticed before, her head, and her hijab are precisely one of the exotic symbols that immediately underlines the difference between her and the viewer (see also Van Leeuwen and Jewitt 2001, 64–96 for a similar analysis). The angle of the picture is not completely at the same level as that of the represented participants but slightly higher, underscoring the viewer's power in the interactive relationship.

The problem with such kinds of representations is that they reproduce, and reinforce, a binary view that constructs the refugee as the 'other' and suggests that as far as we can emotionally relate, their situation is somewhat alien from us. The binary view of 'us' looking at 'them' implies that the suffering and the difficult condition these people are going through takes place elsewhere (Szörényi 2006) and that the problem is coming from the outside (Calhoun 2004). Of course, this image is not unique in this genre. The visual construction of the 'other' is a commonly recurrent theme in the group of images analyzed. Through different combinations of visual social semiotic resources it is present across all four organizations. Although, as I will show in the next chapter, the contemporary preferred humanitarian representation is that of an individual portrayed frontally through a close-up shot and looking straight into the camera, images constructing the 'other' through different levels of saliency given to the represented participants, the oblique (or from behind) angle of the picture, the medium-far distance of the shot, and the specific narrative structure, all constitute an important aesthetic pattern of NGOs visual production.

The 2016 Save the Children's advertising campaign entitled Have we got everything/We've got nothing is paradigmatic. All images are built on the dissociation of one represented participant in two different elements. The one on the left, is clearly 'our' peaceable, flowless, light-hearted world. On the right, the suffering, disordered and dark world of displaced people. There is no indication that the people depicted on the right are Syrians because it is not the point. No details identify the ones on the left as Italians, Americans or Greeks. What is implied is that on the left there is a general representation of holidays (as typically represented in the Global North) while on the right, a general humanitarian victim situation. Moreover, in the reproduction of these dichotomous categories, the contraposition between what Hall (1996) refers to "belongingness" and "otherness" is fixed and naturalized. The value of this composition is quite meaningful. In visual social semiotics there are three semiotic resources at play in the compositional meaning: information value, salience and framing. The former and the latter are particularly relevant here. The respective position of the elements promotes a specific reading according to their placement in the image. In the Western world, accustomed to left-to-right orientation in reading and writing, the elements on the left side represent the "given", that on the right, the "new" (Van Leeuwen and Jewitt 2001). Therefore, not only do these images invite us to a binary view of the world, one clearly demarcated between 'us' and 'them', but also they naturalize 'our' and "'their' world as something given, unproblematic, ordered as opposed to a new, disordered and emergency world. The framing, in turn, splits the image into two parts, accentuating the dichotomous division and disconnection between two worlds that are presented as not belonging together or potentially having any point of contact.

Obviously, the scope of the organization is not to underline otherness and division but rather to point to the immoral contrast between happiness/unhappiness, having plenty of/being deprived of, peace/war. It is important to consider these images within the context of the NGOs' communication production, whose scope is to draw attention to the humanitarian situation and mobilize public support. The problem with this is that they also reproduce dichotomic understanding of us/them, normalcy/abnormality that are not contextualized into specific historical and political dynamics. As Bleiker et al. (2014) have pointed out, images exercise a great power in shaping cultures of hospitality. This is even more true in the case of NGOs' visual representations, that, differently from other kind of representations (media and political discourses for example), certainly have the intention of fostering cultures of welcoming and protection. If we assume, as is fair, that this is probably one of the intentions of these pictures, we should then consider that the representation of the refugee as someone "other" from us, contributes to the production of a discourse of hospitality that is based on distinction, rather than commonalities. This is not problematic because it points out differences among communities, that are of course present and definitively do not need to be hidden. It is rather problematic because it first contrasts with the rhetoric of *after all we are all humans* utilized by humanitarian communication. Even more, it precisely states the opposite, reproducing and reinforcing power inequalities: clearly, there is a 'we' looking at what is happening, required to react to the situation, asked to be in favor of hospitality, and a 'they', looked at, waiting for our action and welcoming, naturalized in a subordinate position of power.

What all these pictures have in common is that they catch the unequal relationship that 'we' and the 'others' have with mobility. As Massey has noted, distinct social groups have different relationships with movement and flows: "some people are more in charge of it than others; some initiate flows and movement, others don't; some are more on the receiving-end of it than others; some are effectively imprisoned by it" (Massey 2010, 149). Pointing out – and implicitly criticizing – this differential power geometry is probably one of the intentions of the NGO when producing such images. However, beside the implicit condemnation of mobility inequalities, what these pictures do is also to reproduce, and worse, naturalize the representation of dichotomic, (and extremely generic) categories of 'we' versus 'others'. In fact, while surely aiming at eliciting indignation for the inequalities between 'our' lucky and peaceful world and 'their' unlucky, dystopian and suffering world, these images offer a simplistic account of the situation, where the causes of the suffering are unclear and eventually irrelevant.

The problem is that these kind of depictions accusing "nobody and everybody'" stimulate a response that is eventually misdirected and unhelpful (Berger 1980, 39). By juxtaposing a 'normal' condition of life in the Global North with the emergency situation of the victims of the Global South, this representation fails to admit the protracted nature of many of these emergency situations, how they interrelate with international politics and eventually their systemic presence in world dynamics. The constitution of the category of the 'other' has, of course, also the strong potential to define the self as an opposed category. The wealthy homogeneous audience of the Global North, moving around the world for purposes of tourism, is an essentialized group as much as the refugees. It would certainly be interesting when thinking about the construction of the self-identity in NGOs' visual production. However, this goes behind the scope of this section. What I find more relevant in the analysis of the visual construction of otherness is how this visual pattern contributes to the constitution of the refugee as a threat. The binary representation securitizes the refugee issue by depicting it as a condition of exception to the normality of life. Refugees' identity is securitized for "successful securitization of an identity involves precisely the capacity to decide on the limits of a given identity, to oppose it to what it is not, to cast this as a relationship of threat or even enmity, and to have this decision and declaration accepted by a relevant group. In the process of dividing between "us" and "them," the concept of societal security echoes the determination of friends and enemies beneath Schmitt's concept of the political, and the acceptance of absolute decision in conditions of emergency" (Williams 2003, 520).

5.3 Conclusion

This chapter has explored the different visual themes used by NGOs that contribute to represent refugees in terms of *threat*, as threatening. While certainly not the dominant visual pattern in aid organizations' visual productions, the images that depict displaced Syrians as threat are neither absent from the humanitarian discourse nor so infrequent as to be ignored. This is particularly relevant if one considers that NGOs do not act in a communication vacuum. On the contrary, they are one of the actors able to present a narrative on the refugee issue among other communication actors, notably populist and right-wing political parties and the xenophobic media that sustains them. Since they have quite consistently represented the European refugee 'crisis' in terms of the exclusionary rhetoric of othering, politics of fear (Krzyżanowski et al. 2018) and securitization (see among others De Genova 2016; Rheindorf and Wodak 2018), one would expect transnational humanitarian NGOs to produce an alternative message. This has, to some extent, happened to with the relief organizations' focus on refugees as threatened victims (the subject of the next chapter). Against this backdrop however, the existence of images resonating with mainstream securitized accounts sits uneasily alongside the intention of humanitarian organizations to challenge such a depiction.

The analysis has empirically shown how transnational humanitarian NGOs contribute to a securitized representation of refugees. I have demonstrated how the pictures depicting refugees' movements through narrative structures of *vectoriality with no goal* have contributed to the representation of displaced Syrians as a threat, as arrows penetrating 'our' tranquil territory. I have also shown how the visual rhetoric of emergency produced by NGOs constitutes the protracted reality of displacement as a disorder to the otherwise normal flow of things that needs to be addressed by extraordinary measures, reproducing and reinforcing a securitized discourse around migration and migration management. By looking specifically at the iconography of the boat, so typical of visual accounting of the 2015–2016 migration 'crisis', I have endeavored to document how relief agencies have failed to challenge and completely move away from media and populist political accounts depicting masses of people arriving on European shores. Despite the introduction of innovative visual elements, NGOs' repertoire on boats and 'boats people' has also continued to represent refugees in a dehumanized and, at the same time, incumbent way, liable to arouse feelings of uncertainty, discomfort and fear. Moreover, I have argued that the large use made by NGOs of different kind of images based on conceptual structure contributes to an essentialized representation of the refugees that takes away their individuality and humanity, even when it represents them in close-up portraits that underline their personal features. This depiction also defines refugees as victims and vulnerable people in need of assistance, eventually creating a contrast between the vulnerable victims that have become the poster image of refugeeness and deserving of our support – typically women and children, older people, families, and people with disabilities – and excludes the group of adult men that constitute at least half of the entire refugee population. Finally, I have shown how the depiction of the refugee as someone alien from 'us' has the effect of reinforcing a simplistic account of the 'other' defined in binary terms as opposed to a generic 'us', and, most importantly, as an unexpected exception from an otherwise peaceful world that contributes to the representation of the refugee as a threat to an allegedly 'normal' order of things.

References

Agamben, G. (2005). *State of exception* (Vol. 2). University of Chicago Press.

Andersson, R. (2012). A game of risk: Boat migration and the business of bordering Europe (Respond to this article at http://www.Therai.Org.Uk/at/Debate). *Anthropology Today, 28*(6), 7–11.

Arnheim, R. (1956). *Art and visual perception: A psychology of the creative eye.* Univ of California Press.

Arnheim, R. (1982). The power of the centre. *A Study of Composition in the Visual Arts, Berkeley.*

Berger, J. (1980). Photographs of agony. *About Looking, 37,* 40.

Bigo, D. (2002). Security and immigration: Toward a critique of the governmentality of unease. *Alternatives, 27*(1_suppl), 63–92.

Bleiker, R., Campbell, D., Hutchison, E., & Nicholson, X. (2013). The visual dehumanisation of refugees. *Australian Journal of Political Science, 48*(4), 398–416.

Bleiker, R., Campbell, D., & Hutchison, E. (2014). Visual cultures of inhospitality. *Peace Review, 26*(2), 192–200.

Butler, J. (2009). *Frames of war.*

Buzan, B., Wæver, O., & De Wilde, J. (1998). *Security: A new framework for analysis.* Lynne Rienner Publishers.

Calhoun, C. (2004). A world of emergencies: Fear, intervention, and the limits of cosmopolitan order. *Canadian Review of Sociology/Revue Canadienne de Sociologie, 41*(4), 373–395.

Calhoun, C. (2010). The idea of emergency: Humanitarian action and global (dis)order. In D. Fassin & M. Pandolfi (Eds.), *Contemporary states of emergency. The politics of military and humanitarian interventions* (pp. 29–58). New York: Zone Books.

Campbell, D. (2007). Geopolitics and visuality: Sighting the Darfur conflict. *Political Geography, 26*(4), 357–382.

Campesi, G. (2011). The Arab spring and the crisis of the European border regime: Manufacturing emergency in the Lampedusa crisis.

CARE. (2015). *The situation is deteriorating rapidly for refugees in the Balkans.* 21 October 2015. https://www.facebook.com/carefans/photos/a.62118502158.68038.30139072158/10153 607535447159/?type=3&theater

Chouliaraki, L. (2013). *The ironic spectator: Solidarity in the age of post-humanitarianism.* Wiley.

Cottle, S., & Nolan, D. (2007). Global humanitarianism and the changin aid-media field: "Everyone was dying for footage". *Journalism Studies, 8*(6), 862–878.

Cuttitta, P. (2017). Repoliticization through search and rescue? Humanitarian NGOs and migration management in the central mediterranean. *Geopolitics,* 1–29.

De Genova, N. (2016). The "crisis" of the European border regime: Towards a Marxist theory of borders. *International Socialism, 150,* 31–54.

DeChaine, D. R. (2002). Humanitarian space and the social imaginary: Médecins Sans Frontières/Doctors without borders and the rhetoric of global community. *Journal of Communication Inquiry, 26*(4), 354–369.

Dijkzeul, D., & Moke, M. (2005). Public communication strategies of international humanitarian organizations. *International Review of the Red Cross, 87*(860), 673–691.

Dillon, M., & Reid, J. (2000). Global governance, liberal peace, and complex emergency. *Alternatives, 25*(1), 117–143.

Duffield, M. (1994). Complex emergencies and the crisis of developmentalism. *IDS Bulletin, 25*(4), 37–45.

Duffield, M. (2001). Governing the borderlands: Decoding the power of aid. *Disasters, 25*(4), 308–320.

Duffield, M. (2007). *Development, security and unending war: Governing the world of peoples*. Polity.

Falk, F. (2010). Invasion, infection, invisibility: An iconology of illegalized immigration. *Images of Illegalized Immigration: Towards a Critical Iconology of Politics*, 83–100.

Fehrenbach, H., & Rodogno, D. (2015). "A horrific photo of a drowned syrian child": Humanitarian photography and NGO media strategies in historical perspective. *International Review of the Red Cross, 97*(900), 1121–1155.

Financial Times. (2016). EU border force flags concerns over charities' Interaction with migrant smugglers. *Financial Times*, 15 December 2016. https://www.ft.com/content/3e6b6450-c1f7-11e6-9bca-2b93a6856354

Friese, H. (2017). Representations of gendered mobility and the tragic border regime in the mediterranean. *Journal of Balkan and Near Eastern Studies*, 1–16.

Furedi, F. (2005). *Politics of fear: beyond left and right*. London: Continuum.

Gale, P. (2004). The refugee crisis and fear: Populist politics and media discourse. *Journal of Sociology, 40*(4), 321–340.

Hall, S. (1996). New ethnicities. *Stuart hall: Critical dialogues in cultural studies*, 441–449.

Hansen, L. (2011). Theorizing the image for security studies: Visual securitization and the muhammad cartoon crisis. *European Journal of International Relations, 17*(1), 51–74.

Hermanin, C. (2017). Immigration policy in Italy: Problems and perspectives. *IAI Working Paper, 17*(35).

Huffington Post. (2016). Volunteers helping refugees in Greece fear government clampdown. *Huffington Post*, 4 February 2016, sec. The WorldPost. https://www.huffingtonpost.com/entry/greece-volunteers-arrested-lesbos_us_56b37bdde4b01d80b2454c6f

Huysmans, J. (1995). Migrants as a security problem: Dangers of "securitizing" societal issues. In *Migration and European integration: The dynamics of inclusion and exclusion*, by Robert Miles and Dietrich Thranhardt. London: Pinter Publishers.

Huysmans, J. (1998). Revisiting Copenhagen: Or, on the creative development of a security studies agenda in Europe. *European Journal of International Relations, 4*(4), 479–505.

Huysmans, J. (2000). The European Union and the securitization of migration. *JCMS: Journal of Common Market Studies, 38*(5), 751–777.

Huysmans, J., & Squire, V. (2009). Migration and security. In *Handbook of security studies* (pp. 169–179). London: Routledge.

IOM. (2017). *Missing migrants project*. 2017. https://missingmigrants.iom.int/

Jewitt, C., & Oyama, R. (2001). Visual meaning: A social semiotic approach. *Handbook of Visual Analysis*, 134–156.

Kress, G. R., & Van Leeuwen, T. (1996). *Reading images: The grammar of visual design*. Psychology Press.

Krzyżanowski, M., Triandafyllidou, A., & Wodak, R. (2018). *The Mediatization and the politicization of the "refugee crisis" in Europe'*.

Kushner, T., & Knox, K. (1999). Refugees in an age of genocide: Global. In *National and local perspectives during the twentieth century*. London: Frank Cass.

Malkki, L. H. (1995). Refugees and exile: From "refugee studies" to the national order of things. *Annual Review of Anthropology, 24*(1), 495–523.

Malkki, L. H. (1996). Speechless emissaries: Refugees, humanitarianism, and dehistoricization. *Cultural Anthropology, 11*(3), 377–404.

Manzo, K. (2008). Imaging humanitarianism: NGO identity and the iconography of childhood. *Antipode, 40*(4), 632–657.

Massey, D.. (2010). *A global sense of place*. Aughty.org.

Massumi, B. (1993). *The politics of everyday fear*. U of Minnesota Press.

Migration Policy Centre. (2016). *Syrian refugees*, 2016. http://syrianrefugees.eu/

MSF. (2015). *Photo of the week: Idomeni migration route, Greece*. Médecins Sans Frontières (MSF) International.2015.http://www.msf.org/en/article/photo-week-idomeni-migration-route-greece

Musarò, P. (2017). Mare nostrum: The visual politics of a military-humanitarian operation in the mediterranean sea. *Media, Culture & Society, 39*(1), 11–28.

Nolan, D., & Mikami, A. (2013). "The things that we have to do": Ethics and instrumentality in humanitarian communication. *Global Media and Communication, 9*(1), 53–70.

Nyers, P. (1999). Emergency or emerging identities? Refugees and transformations in world order. *Millennium, 28*(1), 1–26.

Ogata, S. (1993). *The state of the world's refugees: The challenge of protection*. New York.

Oxfam. (2016a). *Stranded in Greece: The long refugee road to nowhere*. 2016. https://www.oxfam.org/en/refugee-and-migrant-crisis/stranded-greece-long-refugee-road-nowhere

Oxfam. (2016b). The migrants' winter walk by Oxfam international on exposure. *Exposure*, 2016. https://oxfaminternational.exposure.co/the-migrants-winter-walk?locale=encategories%2Ffamilycategories%2Fbusinesscategories%2Ffamilycategories%2Ffamilycategories%2Fwedding scategories%2Fweddingscategoriescategories%2Fbusinesscategories%2Fbusiness%3Fmore% 3Dtrue%3Fmore%3Dtrue?more=true

Oxfam. 2016c. *This is not about refugees. This is about people*. 2016. https://www.facebook.com/oxfamGB/photos/a.96693231395.100148.7214031395/10153583086306396/?type=3.

Pugh, M. (2004). Drowning not waving: Boat people and humanitarianism at sea. *Journal of Refugee Studies, 17*(1), 50–69.

Rheindorf, M., & Wodak, R. (2018). Borders, fences, and limits—protecting Austria from refugees: Metadiscursive negotiation of meaning in the current refugee crisis. *Journal of Immigrant & Refugee Studies, 16*(1–2), 15–38.

Saunders, N. (2014). Paradigm shift or business as usual? An historical reappraisal of the "shift" to securitisation of refugee protection. *Refugee Survey Quarterly, 33*(3), 69–92.

Save the Children. (2015). *Help is on the way for child refugees | save the children international*. 2015. https://web.archive.org/web/20170410085451/https://www.savethechildren.net/article/help-way-child-refugees

Save the Children. (2016). Children on the move in Europe by save the children. *Exposure*, 2016. https://savethechildreninternational.exposure.co/children-on-the-move-in-europe.

Sea Watch. (2016). Sea-watch fears criminalization of civilian rescue forces in the 2017 election year • Sea-watch e.V. *Sea-Watch e.V.* (blog). 27 December 2016. https://sea-watch.org/en/sea-watch-fears-criminalization-of-civilian-rescue-forces-in-the-2017-election-year/

Sontag, S. (1973). *On photography* [Online]. Electronic Edition 2005. New York: RosettaBooks LLC.

Stocchetti, M., & Kukkonen, K. (2011). *Images in use: Towards the critical analysis of visual communication* (Vol. 44). John Benjamins Publishing.

Szörényi, A. (2006). The images speak for themselves? Reading refugee coffee-table books. *Visual Studies, 21*(01), 24–41.

UNHCR. (2016). *Refugees and migrants sea arrivals in Europe - monthly data updates December 2016*. 2016. https://data2.unhcr.org/en/documents/download/53447

Van Leeuwen, T., & Jewitt, C. (2001). *The handbook of visual analysis*. Sage.

van Liempt, I. (2016). *Rescuing refugees in the context of Europe's fight against human smuggling*. Allegra. 5 April 2016. http://allegralaboratory.net/rescuing-refugees-in-the-context-of-europes-fight-against-human-smuggling/

Vaughan-Williams, N. (2015). "We are not animals!" humanitarian border security and zoopolitical spaces in Europe. *Political Geography, 45*, 1–10.

Watson, S. D. (2009). *The securitization of humanitarian migration: Digging moats and sinking boats*. Routledge.

Watson, S. (2011). The 'human' as referent object? Humanitarianism as securitization. *Security Dialogue, 42*(1), 3–20.

Williams, M. C. (2003). Words, images, enemies: Securitization and international politics. *International Studies Quarterly, 47*(4), 511–531.

Wittgenstein, L. (1953). *Philosophical investigations* (GEM Anscombe, Trans.).

Chapter 6
Threatened, the Refugee as the Referent Object

6.1 Introduction – The Refugee as Referent Object

Since its inception, humanitarian communication has consistently represented beneficiaries as referent objects of a threat, as threatened. Images of victims, whether in the traditional representation of a sea of humanity' (Malkki 1996) or in the more recent aesthetic style of the individual portrait, have consistently constituted the large bulk of humanitarian NGOs' visual production. This chapter focuses on the representation of Syrian refugees as 'threatened' to show how this depiction of refugees is just another form of securitization, whereby Syrians are depicted as infantilized and passive victims in need of external intervention. In order to do so, it is worthwhile digressing to understand how remarkable have been the structural changes that humanitarianism has undergone over the last quarter century and how new relief assistance' modalities, while seeking to putting individuals and their rights center stage have also primarily represented them in terms of victimhood.

Since the 1990s, humanitarian actors have devoted increasing attention to the protection of human rights. From the purely needs-based approach of early humanitarianism, where life-saving assistance was delivered to people in danger according to their practical needs, there has been a shift toward a rights-based approach. The overarching objective of humanitarian assistance is today not merely focused on people's lives, but it is now equally interested in protecting people's rights. This laudable aspiration has been realised by a new focus on protection.[1] Despite the intense and heated debate on the necessity for and effectiveness of the rights-based approach (Chandler 2001; Žižek 2005; Brems 1997; Coomaraswamy 1994; Duffield 2014; Fox 2001; Rieff 2003; Slim 2003; DuBois 2007), its diffusion in present day humanitarian action is indisputable.

---

[1] In the humanitarian context, the term protection refers to "humanitarian work [that] extends beyond physical assistance to the protection of a human being in their fullness. This means a concern for a person's safety, dignity and integrity as a human being" (Slim and Bonwick 2006, 30).

A. Massari, *Visual Securitization*, IMISCOE Research Series,
https://doi.org/10.1007/978-3-030-71143-6_6

Since genocide in Rwanda and the former Yugoslavia in the 1990s, the international community has elaborated another concept, one which includes both human rights protection and the right to be protected: that of human security. The term was officially embraced in 2001 by the Commission on Human Security (CHS), a body established in response to the UN Secretary-General's call at the 2000 Millennium Summit for a world "free from want" and "free from fear": "Human security means protecting fundamental freedoms – freedoms that are the essence of life. It means protecting people from critical (severe) and pervasive (widespread) threats and situations. It means using processes that build on people's strengths and aspirations. It means creating political, social, environmental, economic, military and cultural systems that together give people the building blocks of survival, livelihood and dignity" (Ogata and Sen 2003, 4).

The elaboration of the notion of human security has had several important implications. The concept paved the way for development of the Responsibility to Protect (R2P) principle.[2] This provided the moral justification for a series of military interventions – including that in Libya in 2011 which was judged to be a model R2P operation (Kuperman 2013) – and put protection activities at the core of humanitarian action (Slim and Bonwick 2006). Most importantly, the concept of human security generated a theoretical shift in the way international security had hitherto been perceived. Rather than only having States as the focus of international security, people have been included in the picture, at the very center of the international security system (Duffield and Waddell 2006). In this new framework, the international community is impelled to act not only in the case of State aggression against another State, but also when serious violations of human rights are perpetrated against people by another non-state actor and/or the State itself. In the language of securitization theory, the human being becomes in this context the *referent object* of security and the violation of human rights constitute the *threats* from which people need to be protected. In this context, when people are threatened, humanitarian action is a legitimate response strategy to address human security (Watson 2011). Paying attention to the securitizing role of humanitarian NGOs in the human security framework, allows us to unpack the mechanism through which relief agencies "hold a privileged position in the enunciation of human insecurity, in which a reified and monolithic form of humanity is declared, and that supports existing international norms pertaining to the provision of security for humans" (Watson 2011, 5).

---

[2] The R2P principle, promoted by the UN Office on Genocide Prevention and the Responsibility to Protect (https://www.un.org/en/genocideprevention/about-responsibility-to-protect.shtml), has been advanced as a solution to square the circle between the principles of sovereignty and human rights. In his 2000 Millennium Report, the UN Secretary-General Kofi Annan specifically asked: "If humanitarian intervention is, indeed, an unacceptable assault on sovereignty, how should we respond to a Rwanda, to a Srebrenica, to gross and systematic violation of human rights that offend every precept of our common humanity?" The International Commission on Intervention and State Sovereignty (ICISS) was created in 2001 in order to answer that question. In its report, the Commission affirmed that States have the responsibility to protect their citizens, but in case of States' failure to protect its own people, the responsibility is transferred to the international community.

This chapter focuses on NGOs' humanitarian representation of Syrians on the move as *threatened*. The shift of attention from their needs (caused by famine, poverty, or conflict) to threats to their rights is also noticeable at the visual level. NGOs' humanitarian communication. Their attention has been increasingly focused on the human being as referent object of the threat: whether it be violence, war, exploitation or denial of access to basic services. The point is not the threat, but the fact that the represented participants of the images – the actual or potential beneficiaries of humanitarian action – are threatened as their rights are at risk. The typical image of present-day humanitarian crisis is no longer that of a starving baby, but that of a displaced child whose right to education has been taken away by war. Beneficiaries of humanitarian intervention are no longer represented (only) as people in need, but as rights-holders in need of protection. Contrary to what one could expect, this kind of depiction is not particularly linked to negative images more than it is to positive ones. It is, rather, its focus on the incumbent threat faced by the represented participants that links them with the concepts of protection and human security. In the language of securitization theory, the beneficiaries of the humanitarian intervention are visually depicted as referent object of a threat, as *threatened*.

Within the logic of human security and protection, the humanitarian actor's intention is to direct attention toward the persons of concern, and the need to act to prevent the (further) violation of human rights. In an "economy of attention" (Citton 2014) this is one of the way relief agencies in general, and NGOs in particular, compete with media and populist political accounts to offer an alternative account of emergency. Instead of focusing on the *threat* represented by people on the move, humanitarian organizations put the emphasis on the people and the situation of emergency that are *threatening* them.

Inspired by the theoretical framework offered by securitization theory and looking at the depiction of refugees as *referent object of a threat*, this chapter aims to explore how visual representation of the humanitarian subject as threatened contributes to the securitization of people on the move. Through a visual analysis of NGOs photographic material, I argue that several aesthetic patterns, which also happen to be the most recurrent ones in present day humanitarian communication, contribute to an account of displaced Syrians as threatened. Although this depiction is quite evidently intended to challenge dominant discourses of refugees as a security threat, it not only fails to keep the emancipatory promise implied by the rights-based approach to human security, but it also contributes to the reinforcement and the reproduction of a securitized account in which people on the move, as rights-holders yet passive victims, are in need of external protection. Against this background, humanitarian intervention is, therefore, presented as among suitable strategies to address human insecurity.

This chapter outlines the different visual themes that contribute to the depiction of Syrians on the move as threatened. Given the importance that the protection of the humanitarian subject whose rights are (or have been) threatened has in contemporary humanitarianism and its relevance within the human security/right based framework, the following section addresses this relatively new concept of humanitarian protection and the connection that it has with securitization. The chapter proceeds with the presentation of the analysis of the most recurrent visual patterns that emerge from the literature connected with representation of the victims and people in need. It starts by presenting the iconography of pity, putting it into connection with the religious concept of *pietas*[3] and compassion. It then outlines the different aesthetic patterns that are conducive to a victimized depiction of people on the move by taking into account more negative but also more positive types of photographic representation. The chapter continues with the exploration of the over-representation of children in NGOs' communication regarding the 'Syrian refugee crisis' and the ideological meanings connected with the infantilization of the humanitarian subject. In the fourth section, I deal with the photographic representation of physical pain and death to investigate the limits and/or the emancipatory potential of the visual theme of suffering in the framework on relief agencies' visual production. The topic of innocence is the subject of the subsequent section aiming to show how humanitarian depiction of the threatened victim is often complemented by an emphasis on innocence as a necessary characteristic of displaced people in order for them to be considered as deserving beneficiaries of relief aid. The penultimate last section analyzes the visual theme of the humanitarian worker as visual metaphor of the savior hero valiantly intervening to protect the threatened victim. To conclude, I show how these very different kinds of visual representations of the humanitarian subject as threatened contribute to a securitized representation of Syrians on the move.

6.2   Protection and Securitization

Protection is the operationalization of the concern of humanitarians to assist the threatened individual. As, such, it is an overarching concept that encompasses all the visual themes that depict people on the move as referent objects of a threat, as people, that need, exactly, to be protected. Although commonly perceived as a dimension antithetical to the securitarian one, protection and securitization appear not only often intimately connected, but to a certain extent, two sides of the same coin.

Over the last two decades, NGOs have paid increasingly attention to protection activities. This growing sector of intervention has been added to the traditional emergency operations around health, food assistance, education, water and

---

[3] *Pietas* is an "abstract divinity of the Romans, which expresses the set of duties that man has both towards men in general and towards parents in particular." In the modern sense, the term refers to the feeling of affectionate pain, of moving and intense participation towards those who suffer.

sanitation, non-food items, and shelter. In the ALNAP Protection Guidelines,[4] Slim and Bonwick define it as "humanitarian work [that] extends beyond physical assistance to the protection of a human being in their fullness. This means a concern for a person's safety, dignity and integrity as a human being" (Slim and Bonwick 2006, 30). Humanitarian protection is, basically, the operationalization of the rights-based approach, the way through which relief agencies (whether governmental or non-governmental) translate into practice the aspiration to address human (in-)security.

Despite the very noble scope of protection, its implementation has several problematic implications. Schubert (2007) has shown how protection, is fundamentally a political act, given its innate struggle to legitimize who has the authority to protect and within which limits. Although presented as an apolitical and ethical intervention, it is not only intrinsically political, but it also shares a very similar logic with security discourse in legitimizing an expert role acting on behalf of the 'other' lacking political agency. In some cases, the language of protection has been used to implement activities that have clearly contrasted with the intention of the people they meant to protect. Musarò (2013) has pointed out how Frontex's action in preventing irregular migration has been presented in terms of humanitarian intervention to save lives and protect the victims of smuggling, while, in reality, the agency's primary goal was and is to prevent migrants from reaching Europe and to forcibly return as many as possible to their countries of origin.

It is therefore not ironic that a UK flagged ship, which I saw during fieldwork in Lesvos anchored in Mitilene, just outside my hotel displayed both languages: signage saying 'Border Force' juxtaposed with that of 'Protector'. If those examples are about a security agent that claims a protection role, the interplay of these two seemingly conflicting yet interconnected domains occurs also the other way around. A child protection officer working for a local Greek NGO on Lesvos shared her concerns. Talking about their work with minors who are registered as adults,[5] she commented "that is what we do, we are the good police in there. Our work is not child protection, why do they call it protection? We are there basically protecting the system, not the children."[6] Similarly, a Save the Children child protection manager expressed her frustration following a meeting the child protection unit in Moria Camp: "the other day we met with other child protection actors in Moria to discuss measures to protect from violence and abuses the minors registered as adults in the camp and the only solution we could think of was to build other fences. To protect them we basically had to put them in a prison inside a prison."[7]

---

[4] ALNAP is "a global network of NGOs, UN agencies, members of the Red Cross/Crescent Movement, donors, academics and consultants dedicated to learning how to improve response to humanitarian crises" (see: https://www.alnap.org/)

[5] The registration of minors as adults, due to contested age-assessment procedures, was at the time of research a key protection concern. For more information see: Achilli et al. 2017

[6] Interview with a Praxis child protection officer, Mitilene, Greece, 5th May 2017.

[7] Interview with a Save the Children child protection manager, Mitilene, Greece, 4th May 2017.

This complex interplay is not surprising. Protection and securitization alike require the disconnection of an issue from its political dimension for its framing in terms of emergency. In both cases the political context and the individual or collective political claims of the people involved are overlooked in the name of a greater priority. Moreover, as Carling and Hernández-Carretero (2011) have shown, the narrative of migrants' protection simultaneously works as a policy objective and powerful rhetoric to legitimize control measures. "Images of destitute and vulnerable migrants – which can seem at odds with the narrowly security-oriented narrative – underpin the policy narrative of protection. Because ostensibly protective intervention is often compatible with control measures, the narrative of protection achieves a sense of closure" (Carling and Hernández-Carretero 2011, 45). In this sense, protection and securitization appear as two sides of the same coin. The visual analysis of humanitarian NGOs' representation of people on the move starts from this assumption and looks at the different aesthetic patterns that depict displaced Syrians as threatened, underlying different aspects of this interconnection between protection and securitization.

There is another dynamic of protection and its securitizing role toward the threatened victim that has emerged from the visual analysis. NGOs' attempts to show individuals' traumatic experiences, clearly observable in the humanitarian photographic account of the refugee crisis, confirms the argument advanced by Pupavac (2001b) about psycho-social activities. She argues that such trendy interventions, increasingly at the core of emergency protection programmes in the field, have come to constitute a "new form of international governance based on social risk management" (Pupavac 2001b, 358). This governance, that she calls therapeutic, has permeated relief strategy. Humanitarian interventions are based on psychosocial risk management and have a double effect. On the one hand they conceal or hinder local coping strategies to displacement and violence. On the other they pathologize war-affected individuals, considering them as either being 'in recovery' or 'in remission', never recovered, "but ever after haunted by their trauma and at risk of being re-traumatised by their memories" (Pupavac 2001b, 264). We can see a gradual shift from traditional humanitarian pictures of masses of undistinguished individuals to the preference for portraits of individuals. In the representation of the humanitarian subject as threatened, the traumatic experience is assumed and depicted inside the individual, and the threat is internalized in the body of the victim. In this way, people on the move, intrinsically threatened, visually results in identification of the need for therapeutic governance. However, as Pupavac (2001b) has highlighted, several studies on the assumption of trauma in conflict-affected societies have challenged the idea that this assumption can be made. For appearances of clinical conditions are particular rather than universal and specific factors mediate how individual experience war.

6.3   Visually Threatened – Human Security and Securitization

6.3.1   Pity (Fig. 6.1)

A mother with a suffering child is probably one of the most powerful images of pity, intended as the feelings of painful and thoughtful participation in the unhappiness of others.[8] In this picture a little girl, clearly in distress, is lying on her mother's lap. We can feel the mother's presence through the warm and protective embrace of her hands, and her involvement through the gaze turned to her daughter, but we cannot see her face. Her head is cut off from the frame. The little girl, instead, at the center of the scene, is looking straight at us, thus establishing a direct contact with us. She is the focal point of the image as everything else is either cut out or left out of focus. The eye contact is a typical feature of 'demand images' where the gaze of the represented participant demands something from the viewer, to enter into some kind of abstract relationship (Kress and Van Leeuwen 1996). The kind of relationship expected in this picture is revealed by the little girl's expression. Her suffering look, combined with the close-up distance and frontality of the shot, demands empathy and compassion. The text accompanying the picture informs us that the little girl is

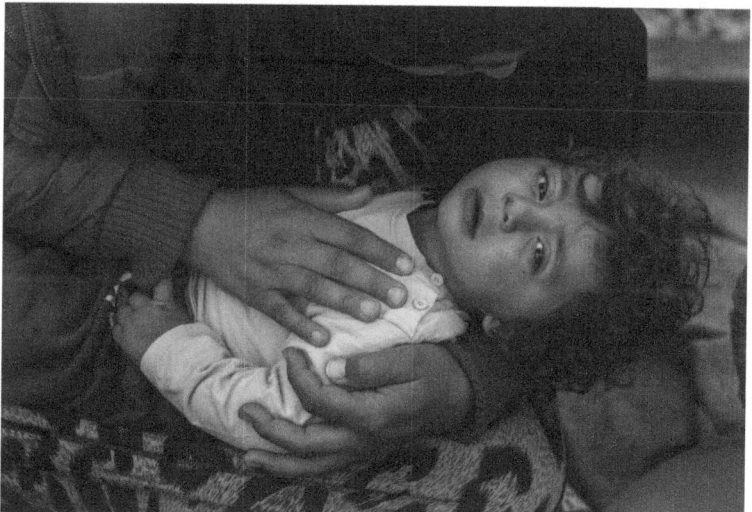

Fig. 6.1 ©Pedro Armestre/Save the children

---

[8]The term *pity* is very similar to the term *compassion* although in religious terms pity is only an emotion and compassion is both an emotion and a virtue. However, the two notions are often used interchangeably and in relation to the religious term of pietas. The terms *pity* and *compassion* differ substantially from the more secular notion of *solidarity* that implies a relationship between an individual and a community while the former are more linked to a one-to-one relation.

sick and cannot stop crying. She and her family are stranded in Greece and waiting to go to Germany to access appropriate care (Save the Children 2016a). The image leaves little doubt about the kind of reaction which is expected. This request for emotional involvement, however, comes also with the suggestion of a specific power relation between the represented participant and the viewer. The high angle of the indicates the viewer's position of power toward the subject of the image (Van Leeuwen and Jewitt 2001), mitigating, in a certain way, the feeling of close social distance and involvement proposed by the other semiotic resources at play. Most importantly, views from above "tend to diminish the individual, to flatten him morally by reducing him to ground level, to render him as caught in an insurmountable determinism" (Martin 1985, 38).

This image powerfully recalls images of *pietas*, an iconographical theme widely present in pictorial art, represented par excellence in religious art of the Renaissance, but also present in secular illustrations. The visual pattern connecting humanitarian representations of pity to the iconography of religious *pietas* is no coincidence: the aesthetic theme of piety, or its more secular synonym of pity, arouses precisely feelings of compassion and empathy for the suffering other. This is probably one of the reasons why pity has been historically one of the most recurrent themes of humanitarian communication.

One of the intentions of this kind of representation is certainly that of showing the suffering of people and soliciting the viewer's emotional involvement. For NGOs such communication is desirable, helping to obtain financial support for the organization's humanitarian work. For relief agencies, one of the key communication objectives is also that of raising awareness on the situation about which the public could not be fully informed. As a Save the Children communications staffer told me: "Currently one of the key messages is around the EU-Turkey deal, on the impact that the deal has on the kids especially on the Greek islands. Their psychosocial status is increasingly worse with more and more kids attempting suicide and arming themselves. We are calling for the EU to improve the condition on the islands, taking family and children out of the detention centers, give them a fair right to asylum and that every European country does it fair share for resettlement".[9]

In the specific case of Save the Children picture above, the image is part of a photojournalistic reportage entitled *Children on the move in Europe*. It clearly sets out its key message: the EU – Turkey Deal that had come into force a few months previously is leaving children (and their families) stranded in Greece, at risk of inhumane conditions and detention (Save the Children 2016a). The message is quite clear and its resonates with discourses of social justice, humanity, and solidarity: there is a humanitarian situation going on in the Greek islands and the organization wants to inform, raise public awareness, and advocate with European Governments

---

[9] Interview with a Save the Children Information, Communication and Media Manager, Athens, 24th March 2017.

for safe and legal routes to Europe and increased resettlement options (see *Our Call for European Leaders* section in Save the Children 2016a).

What I find problematic is that the cover picture presenting the section dedicated to Syrian displacement is instead focusing on pity. The problem with this kind of framing – a framing present, to different degrees, across the four selected NGOs – is that, as Hannah Arendt (1990, 7) has argued, the politics of pity operate under the assumption of the existence of two groups: the unlucky ones who suffer and the lucky ones, who do not suffer but get to observe the suffering. In Arendt's perspective, pity could be considered a step forward compared to compassion. According to the philosopher, compassion operates outside the political space for it is a passion incapable of generalization, one that cannot be applied to an entire class of people, but only to a specific case that stirs the observer and creates co-suffering. Pity, instead, is able to make generalization, to address classes of people, but only in relation to their fortune and misfortune. Pity is different from solidarity, a principle to guide action which takes into account differences in strength and weakness, richness and poverty. Pity, however, only looks at luck and completely disregards justice. For this reason, the politics of pity is also different from the politics of justice. Although pity may have inspired action for the alleviation of suffering, it is based on a discourse of grand emotions, rather than socio-political justice (Chouliaraki 2010). Drawing on Arendt's reasoning, Boltanski (1999) has noticed that, in fact, "for a politics of pity, the urgency of the action needing to be taken to bring an end to the suffering invoked always prevails over considerations of justice" (Boltanski 1999, 5).

There is something else in the iconography of pity that I find interesting. The large majority of pictures of pity examined for this study represented women and children. This is particularly relevant if we consider that the preliminary content analysis did not identify an over representation of women versus men. While children were the subject of the majority of pictures, when it came to adults the two genders were almost equally represented in photos of single individuals (71 women versus 60 men) in photos of group of people (39 groups of women and 44 groups of men). However, in the images of pity women and children were clearly predominant. This is strikingly e at odds with the demographic data on Syrian displacement in Europe in 2015. According to UN data, in fact 62% were men, 16% women and 22% children (Time 2015). Although in the Middle East by the autumn of 2016 the percentage of minors was higher (47%), the entire population was broadly split between the two genders (UNHCR 2016).

Since the visual theme of children is particularly dominant in humanitarian representation and entails specific implications, it will be treated in detail in a dedicated section (see Sect. 6.3.3). Here, I would like to focus on the over-representation of women in images of pity.

In recent years there has been increased attention to gender in the humanitarian sector. Almost all relief organizations (both governmental and non-governmental) have progressively and inexorably integrated gender mainstreaming – the inclusion of a gender perspective into humanitarian policies, practices and project implementation – into their emergency programmes. Although in theory a gender perspective is not particularly focused on women more than on men, the visual presence of women compared to man is overwhelming. This has probably something to do with the fact the humanitarian discourse has often considered vulnerability as essentially situated in the female body, reinforcing a representation of women as victims and passive recipients of aid (Aoláin 2011). The problem with the images of pity predominantly embodied by women is that they reproduce an essentialist depiction of women as intrinsically vulnerable. The over-representation of women in images of pity also resonates with a growing body of scholarship that has critically addressed the narrative of rescuing women and children that has permeated discourses for the legitimization of military humanitarian intervention and humanitarian action (Heck and Schlag 2013; Rosenberg 2002). Most NGOs may have, albeit belatedly, acknowledged that an attention to gender means an attention to all genders and gender dynamics, and not to a specific gender (IASC 2006). However, such a recognition is completely absent at the visual level in images of pity.

6.3.2   Victimization (Figs. 6.2, 6.3)

A little girl, barefoot and wearing an over-large T-shirt, is rubbing her eyes, probably crying, miserable beside a tent. There are no other details in the picture except for the dirty and muddy soil. A close-up portrait where everything is in sharp focus and the girl (and her affliction) are the sole protagonist of the picture. This image – beside the unease potentially provoked in me every time I look at it that I have intruded in an intimate moment – tells us very little about the circumstances that led her to this situation. We do not know why she is crying, or if anyone, apart from us,

Fig. 6.2 ©CARE, Syria Response 20 September 2015

Fig. 6.3 ©CARE, Syria
Response 24 July 2015

is seeing her and will go to comfort her. The photographer has put him/herself (and the viewer thereof) at the same height as the child in an effort to mitigate the asymmetrical relation of power between the observer and the observed. However, in spite of this attempt, the oblique point of view unveils the detachment and moderates our involvement with her feelings: despite sympathy we can have for her, the diagonal angle depicts her as part of another world, not ours. As Krees and van Leeuven (1996) have pointed out, this does not mean that the viewer is forced to accept the specific view suggested by the semiotic resource of the diagonal angle (that of social distance). Yet, before rejecting its meaning, it is necessary to acknowledge the value of this particular semiotic resource.

The post accompanying this image on Facebook specifies the context: "This little girl and her family only wanted refuge. Instead, she got teargas. Help Syrian refugees, their lives & dignity matter." Together with the hashtags #Refugees #EU #SyrianRefugees #RefugeeCrisis, it also indicates a link. I immediately clicked, looking for more information on what had happened: why teargas, who had done that and why? Nothing, for the link does not lead to any further information, only to a page urging donations to help the organization "fight poverty and provide emergency relief" (CARE 2016b). I followed the clue of the teargas easily learning that 3 days prior to publication of the photo the Hungarian police, guarding the border with Serbia, had fired upon people on the move with tear gas and water cannons (The Independent 2015). However, none of this is visible in the picture. There is no border, no police in riot gear, nothing that could give us the sense of what is going on and why the little girl is crying. Once again, we note how the absence of any kind of information on the context reproduces a victimized account of Syrian refugees that conceals the relevance of their socio-political situation.

The representation of people on the move as victims is one of the most discussed visual tropes in the literature. A large body of scholarship has investigated how certain ways of depicting refugees or aid recipients tends to victimize them by describing the represented participants as helpless and passive (Calain 2012; Fassin 2007; Kleinman and Kleinman 1996; Malkki 1996; Rajaram 2002). Overall, the analysis has revealed that all four NGOs publish and produce pictures of victimhood. However, it is important to note that CARE and Save the Children utilize this visual theme to a much greater extent than Oxfam and MSF. The iconography of victimization is different from pity as the former places the emphasis on passivity rather than compassion. In fact, this representation does not necessarily rely on negative images where the subject of the picture is clearly in distress. It can, on the contrary, depict people with a proud look or even smiling. Similarly, passivity and victimization are no longer necessarily associated with images of group of indistinguishable people as a "sea of humanity" (Malkki 1996, 377).

No matter the subject, or the positiveness/negativeness of the image, what all these pictures have in common is that they show the represented participants as passive victims of circumstances. All this is in striking contrast with reassurances given me by a regional communications officer for the Syria response when explaining what a good' picture should look like according to his organization: "We always try to create empowering images, we do not take pictures from certain angles that would make them vulnerable".[10]

However, images of this kind represent displaced people in terms of loss and helplessness. It is not about a specific semiotic resource at play, but, rather, how in each specific picture the combination of various semiotic resources contributes to the depiction of victimhood. It could be the lighting and the depth of field focusing on the situation of misery or the medium size of the frame, which combined with a high and oblique angle detaches the viewer from the represented participant who sits on her hands completely helpless. Despite, sometimes, the strong and proud look of the people, and that some images are taken from a frontal, inherently more equal, angle, it is loss and inactiveness which usually permeate such types of representation.

While the visual topic of the victim is quite widely used in present-day contemporary humanitarian representation, it is also crucial to acknowledge that NGOs, in contemporary depictions of the threatened individual, are also putting a great deal of effort into avoiding images of passive victims in favor of a depiction of people

---

[10] Interview with a CARE Regional Communications Officer for the Syria Response, Amman, 23rd March 2017.

with agency and in need of empowerment. This is true across the four relief organizations. This important change of perspective it is not only stated in several policy documents and guidelines, such as the *NGOs Code of Conduct on Images and Messages* (Concord 2006) or the different internal guidelines that communication practitioners referred to during my field interviews, but it is also quite evident at the visual level. Figure 6.3 shows a boy holding a piece of cardboard with the first letters of the Arabic alphabet. His proud gaze is directed at us as he poses in front of the camera. Behind him, the background in focus situates the picture in the typical semi-arid landscape of the Informal Tented Settlements (ITSs) in which many displaced Syrians live in Syria, Lebanon and Jordan. This is not a detail, as they are considered particularly vulnerable. As a UNICEF/REACH Report explains, "for Syrian refugees who are unable or unwilling to reside in formally established refugee camps and are unable to afford regular housing solutions within host communities, Informal Tented Settlements (ITS) have become the default option, notably for the most vulnerable and impoverished displaced Syrian households" (UNICEF/REACH 2014, 6). The Facebook post accompanying the picture clarifies: "his boy has been displaced with his family members from their home. They are now living in an informal settlement in Syria. The boy is carrying a sign with the Arabic alphabet written on it, expressing that he has lost his education and that he needs to go back to school." The structure of the picture is based on a symbolic attributive process whereby the cardboard – made salient through the central position in the frame, conspicuous color and its display position – stands to signify its symbolic value: the boy's threatened right, the right to education that has been taken away by war and displacement.

This picture, putting the threatened right at the centre of the attention, through its symbolic attributive structure, is paradigmatic of the relatively new rights-based approach of humanitarianism. The image still presents some elements of the old visual perspective. The pose seems quite artificial since it is hardly the spontaneous idea of a little child to create such a symbolic and powerful image. Moreover, the camera is still shooting from above, from a position of power, one of the humanitarian actors supposed to, and poised to, act to defend the boy's rights.

However, the boy is holding some cardboard and is not looking at us with a suffering or demanding expression, but, rather, with a proud look. He has a message. I initially thought that the picture wanted to show how families were coping with the difficult condition of displacement. In Syria, the educational system has been severely disrupted by the ongoing level of violence, and, as I learned during my fieldwork, in Jordan and Lebanon access to education for displaced Syrian children in general, and for those living in ITSs in particular, was problematic for numerous reasons.[11] However, we do not know if this is the message of the child, if this is his priority or it is the only learning resource he has in the harsh conditions in which he is leaving. Since the boy is clearly posing, and the text says that the cardboard is a

---

[11] For more information on this topic, see: Human Rights Watch 2016a, b.

symbol of its lost right to education as any other displaced child, as any other 'universal victim' (Malkki 1996), his individuality, together with his agency and empowerment are diluted. In this case, empowerment and dignity would have been probably best represented by visually documenting the several attempts that Syrian refugees put into self-organizing education activities in the ITSs and the various obstacles to access their right of education.[12]

Undoubtedly, however, looking more generally at this new type of images, it is clearly evident this kind of representation is completely different from that of the passive victim. It is crucial, therefore, to recognize that images of this kind constitute a drastic change from the typical image of the victimized threatened victim of traditional humanitarian communication. In most of the photos of deprived rights-holder victims, the point of view is frontal, and the camera is put at the same eye level as the subject. Women and girls are no longer the predominant subject. On the contrary, the represented participants are mostly boys, in striking contrast with the photos of pity and passive victims. Faces are in full light and expressions are mostly jaunty and confident. They are not passive victims anymore, they are rights-holders. Yet, minors are the main subject of representation. It is as if for the represented participant to be visually *threatened,* s/he has to embody some kind of vulnerability. Although this representation entails a drastic change compared to the helplessness of more 'traditional' visual themes, it still depicts the people on the move as embodying the need for (humanitarian) protection. They are not completely passive victims anymore, but they remain (mostly) people in need of external intervention.

The new, rather more positive and undoubtedly improved, aesthetic pattern of rights-holder subjects is therefore not without some limitation. Although the passive victim has been replaced by the rights-holder, humanitarian intervention is depicted as natural and inevitable to address the wrong and as the only actor able to act to defend the person' rights. The rights-holder representation works with what I describe as a 'Trojan horse effect'. Deceptively, at first glance it seems to deconstruct traditional representation of people on the move as helpless victims but eventually reproduces a discourse in which the rights-holder is portrayed with limited agency: s/he is threatened as his/her rights are threatened and this needs external intervention for protection. Unlike the mythic wooden horse deployed as a stratagem by Odysseus, with this new genre of picture NGOs are making an effort to f challenge the imaginary of the passive victim with no intention to trick the viewer. Nonetheless, the polysemic feature of photography allows for both meanings to be

---

[12] In 2015, while serving as Head of Mission in Jordan of a humanitarian NGO active in the emergency response to the Syrian crisis, I was a member of the Education working group where the topic of informal education was discussed and coordinated with UN agencies and Ministry of Education representatives. Among the Syrian refugees there were several teachers who, amidst the situation of displacement and ban on formal employment in Jordan, organized small schools in the informal settlements. The Government of Jordan asked the NGOs to harmonize and coordinate the emergency education response while at the same prohibiting Syrian teachers from providing educational services. Other obstacles to accessing education for Syrian refugees living in the ITSs have included transportation costs to and from school and lack of required documentation.

present. While putting the individual and his/her rights at the center of attention, it simultaneously reproduces a representation of the rights-holder with very limited agency.

6.3.3   Infantilization

The over-representation of childhood in NGOs humanitarian communication constitutes a particular form of victimization. Among the more than a thousand images collected for this study, two thirds of all images of individuals portray children. Over 60% of group images include minors. Although only one of the four selected organizations has a specific mandated child focus (Save the Children), the significant recurrence of images of minors is noticeable across agencies and particularly evident in the case of SCF 74% of whose images are of children, CARE (38%), Oxfam (32%), and to a much lesser extent MSF (15%). In Chap. 4 I have shown how the use of conceptual structures has the effect of essentializing and de-individualizing individuals' diverse experience of displacement. Here, I would like to focus on the discussion of the predominant recurrence of childhood iconography and its implications in relation to the humanitarian representation of people on the move. Unlike other visual themes analyzed so far, the representation of children in present-day NGOs communication is quite variegated. Portrayals of children in situations of distress, need, suffering, and misery, coexist with equal numbers of images of them happy, playing, holding a favorite toy. Children are represented in groups or alone, in the arms of their parents or standing out in the middle of a mass of adults.

The representation of children and its iconographic value in humanitarian communication has been extensively explored by the literature. Scholars have shown how not only children (and women as we have seen the previous section) embody a particular kind of powerlessness, but also that they rarely look threatening (Malkki 1995, 1996), symbolizing the humanitarian subject by excellence (Bleiker et al. 2014; Burman 1994; Byrne and Baden 1995). By focusing particularly on NGOs' communication, Manzo has argued that "the iconography of childhood is a signifier of an NGO corporate identity" (Manzo 2008, 635). The way organizations represent children works similarly to a brand logic in which agencies encode their humanitarian principles, vision and ideals. Through their use of a dominant iconography of childhood, NGOs shape, interpret and undermine these principles and values producing a representation of children in which multiple and opposite meanings can easily coexist. Manzo has also shown how the humanitarian depiction of minors resonates with colonial ideology, particularly when analysing analogies between the iconography of childhood and that of savagery, both based on the assumption of innocence, dependence and protection. From a psychological perspective, scholars have explained how images of children evoke in adults a sense of power and confidence in their own abilities (Holland 1992) and are utilized to symbolize grownups' concerns about protection and security (Myers quoted in Burman 1994). The iconography of childhood therefore reinforces a colonial and paternalistic account where the (lonely) child, symbol of all victims, needs protection and transnational western NGOs work in loco parentis (Burman 1994). Pupavac has shown how the current humanitarian emphasis on children's rights "has become key to human rights-based international security strategies" (Pupavac 2001a, 95). Present-day child-targeted psycho-social rehabilitation programs, she acutely argued, are the way through which children rights have been operationalized in the humanitarian sector. Children have become crucial to the creation and creation of a "new international ethical order" and key actors to promote social change (*ibid.*, 95).

Unlike the visual themes hitherto analyzed in the case of the iconography of childhood it is not a matter of the way children are depicted, but, rather, the recurrence of the subject, and what this reveals about NGOs' understanding of the population they work to assist. Both the over-representation of children and the focus on child protection activities confirm the paternalistic account of humanitarian action and power relations between transnational Global North-based aid agencies and beneficiaries in the Global South as pointed out in the literature (see, inter alia, Manzo 2008; Pupavac 2001a, 2005). Moreover, childhood, which is "characterized by protection and freedom from responsibilities" (Burman 1994, 3), reproduces and reinforces a discourse of passivation of the victim.

As already noted in relation to the iconography of victimization there has been an evident change in the way children are represented. Although images of starving babies are no longer a recurrent visual theme, the new images of children as rights-holders fails to escape a representation of the recipients of aid as passive, powerless and vulnerable threatened subjects in need of external protection. Likewise, the variety of aesthetic patterns portraying children (such as single children, those attending protection programmes or groups playing) follows the same logic. For all the evident effort by NGOs to communicate more positive images which present children with dignity, and sense of empowerment, the reality remains that over-representation of childhood continue to conceal more than reveal by providing an infantilized representation of people on the move.

Regardless of whether or not children are represented as passive victims or rights holders, the point is that the figure of the child is used as a metonymy[13] of all displaced Syrians in the visual production of NGOs. Since children are in the Western narrative in need of protection and parental care, the predominant recurrence of this visual pattern justifies and reinforces a paternalistic perspective. In this sense, the over-representation of children in humanitarian communication contributes to a depiction of people on the move as infantilized, because mainly illustrated through the visual theme of the minor. This infantilization is problematic because it transfers to the wider population (all refugees) the features normally associated with childhood: vulnerability, need of external (adult) protection, and limited agency.

---

[13] Metonimy is a figure of speech. It "has the effect of creating concrete and vivid images in place of generalities, as in the substitution of a specific 'grave' for the abstraction 'death'" (Encyclopaedia Britannica 2018).

The visual analysis has not only confirmed what Pupavac (2001a) has noticed, that the child as rights-holder is denied the moral agency to act in his/her interest, but has also underlined the not too implicit statement that NGOs are basically the only actor able to intervene to address the wrong and protect child rights. This is particularly evident if we compare the pictures of NGOs' activities with children and the images of parents with their children. While the latter often do not show the traits of the parents (either photographed from the back or simply cut out of the picture), when they 'are in the frame' – as it is indeed the case in many of the images of the selected four NGOs – they are mostly represented carrying or embracing their children. As much as this is a clear visual signifier of the love parents nurture for their own kids, there are almost no images in which parents are engaged in some of their 'natural' parental activities: teaching, playing, feeding, trying to relieve the kids from the traumatic experiences they went through. Whereas all this is completely visually absent from the images of fathers and mother with their children, on the contrary images of NGOs' projects clearly show the organizations to be very active in these activities. In this securitized representation of people on the move, where Syrian refugees are represented trough the metonymy of children, NGOs contribute to a securitized depiction in which an infantilized subject with limited agency is in need of (adult, or in other words top-down) protection.

6.3.4   Suffering, Physical Pain and Death (Fig. 6.4)

The suffering body, especially of children, has been for a long time a dominant visual theme of humanitarian communication (Fehrenbach and Rodogno 2015). The level of violence, physical abuse and death of the Syrian war, and the displacement it has caused, have produced a series of dramatically vivid images of physical pain and death that are difficult to forget. While there are few very strong and powerful pictures representing death – and particularly the death occurring along the Mediterranean migration journey through open symbols such as coffins, lifejackets, and abandoned children' shoes on the shore – in this section I would like to focus on images of physical pain and dead bodies.

Pictures of this kind, published by the four selected NGOs – over 70% of those analyzes for this study were produced by MSF – are mostly close-up or medium distance shots of children in serious pain due to the wounds caused by bombs or chemical weapons or the effects of famine in besieged areas of Syria. In some other cases the bodies are abandoned on a bed or in the arms of someone rescuing them from destroyed buildings. We do not get to know what happened to the people portrayed. Moreover, there is at least one picture showing explicitly a dead person, probably a child. It portrays a mourning child weeping over a dead body and it is part of a photo collage including the following text: "Fewer people died in the first month of Syria's ceasefire than at almost any other time since the war began. US+RUSSIA: SAVE HOPE.SAVE LIVES.SAVE THE CEASEFIRE. Over the past

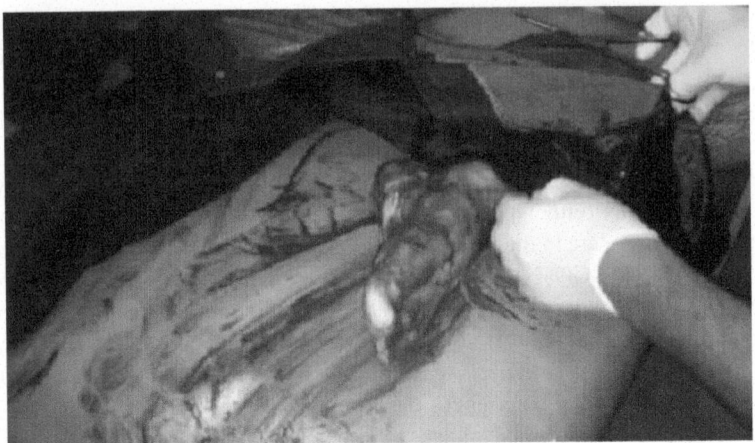

Fig. 6.4 ©MSF, 2013

48 h one Syrian has been killed every 25 min and injured every 13. This is one of them" (CARE 2016a).

In order not to involuntarily fall into the trap of a "pornography of violence" (Bourgois 2001), I have decided to reproduce only one image of suffering that I personally find less shocking than most of its kind within the corpus of photos gathered for this study. Although this image is far from being paradigmatic of the most upsetting ones, it still allows me to make my point without reproducing pictures of extreme suffering and death.

MSF picture shows a foot covered in blood. White plastic gloves are covering the hands that are cutting off the victim's trousers to enable medical intervention. The bloody streaks on the white operating table give a sense of the fast-moving emergency environment of urgent medical assistance. The gloved hands (actor of the image) with the scalpel (vector) are an open symbol of the medical profession. We know – without the need for more details – that there is an aid worker, either a doctor or a nurse, working on the wounded body (the goal of the action). Through its "psychological salience" (Kress and Van Leeuwen 1996, 63) blood dominates the scene. This close-up shot puts the viewer (us) in the middle of the medical emergency. All we can do is to observe and acknowledge the physical pain of the body.

There is something about the representation of suffering and death produced by NGOs that is connected with relief organizations' role to inform the general public what is happening in the field, raise awareness and mobilize support. The underlying idea is that often the victims of violence need their plights to be known to the world and their pain to be visible in order to raise awareness around forgotten crises. As a large swathe of scholarship working on images of atrocities has pointed out, invisibility is not helpful to victims of violence (Sontag 2003; Kleinman and Kleinman 1996; Butler 2004). For many transnational NGOs and certainly the four studied here 'making people' voices heard' is a core priority of their strategies (Save the Children 2016b, 3; Oxfam 2013, 5; Care International 2017, 7). MSF has, as noted earlier, put *témoignage* (bearing witness) at the heart of its humanitarian action. For MSF the principles of impartiality and denunciation, Calain (2012) has observed how humanitarian representation of suffering bodies does not meet the ethical standards usually used in medical practice. In fact, in order to draw attention to a neglected crisis or amidst particularly dramatic events, NGOs sometimes decide to put aside aspirations to use more positive images disseminate images that show extreme suffering and physical pain. When I asked an MSF communication officer when asked what made a 'good' and a 'bad' picture replied: "a good picture would be a picture of a patient or of people receiving MSF care, our activities. The focus is the patient. Since their voices cannot be heard, we try. I do not use communication to promote MSF and generally avoid using pictures with too much blood because I also do not want to hurt the eye of the viewer".[14]

On the role of dramatic pictures of pain, an MSF communication manager expressed her opinion that: "the role of images was much more powerful at the time MSF was founded. Back then, it was extremely important in Ethiopia to photograph what humanitarian workers were seeing in order to inform people who did not know about what was going on. Now that we are bombarded by images sometimes I wonder if they still make a difference. For example, showing the body of Aylan has helped to raise awareness?[15] Surely, but for how long? Its effect has been quick."[16]

However, despite the organizations' internal debate, and at odds with their visual communication guidelines (Concord 2006; Manzo 2008) images of suffering and physical pain are undeniably part of humanitarian NGOs' communication, to such an extent that Denis Kennedy has argued that relief organizations are involved in a "veritable commodification of suffering"(Kennedy 2009, 1).

Several considerations arise. The first has to do with the recurrence of such images. Without denying the importance of making visible the level of atrocities, the quantity of such powerful images contributes to produce what Moeller (1999) calls "compassion fatigue", a resistance to shocking images that eventually dilutes the reason they were taken and distributed in the first place. Since expert actors – the transnational humanitarian NGOs in our case – as Aradau (2004) has rightly pointed out, play a key role in shaping the observers' sensibility, imagination and emotional response, the way suffering is visually represented also defines our understanding of it. In this sense, relief organizations bear a great responsibility for the shaping of the socially constructed aspect of pain. However, because of the impossibility for the viewer to feel what the photographed subject is experiencing, what these images do eventually is to "fetishize pain in their drive to make visible what is essentially unimaginable (…) The body in pain is thus reproduced as an aesthetic visual image, a symbolic icon that stands in for itself as the referent object of political violence" (Dauphinée 2007, 140). The problem is that this representation of the unimaginable

---

[14] Interview with MSF Communication Officer, Amman, 18th May 2017.

[15] She was referring to 3-year-old Alan Kurdi, whose body was found lifeless on the shores of Turkey after boat on which he was travelling with his family capsized.

[16] Interview with MSF Communication Manager, Athens, 24th April 2017.

does not move us closer to the subject, as we cannot physically experience that pain. Sontag concisely pointed out that "to suffer is one thing; another thing is living with the photographed images of suffering" (Sontag 1973, 15). Since we are can only observe the visual manifestation of the pain, visual representation of suffering contributes to underlining the asymmetry between the observer and the body in pain (Chouliaraki 2013; Boltanski 1999).

Even the effort to give suffering a face, demonstrated by the predominant representation of people in pain through close-ups, is severely hindered by the fact that, as Dauphinée (2007) has noticed, the people depicted stand there as iconic representations of the cause of that pain and their individual suffering is completely erased. The representation of physical suffering underlines the impossibility of empathy and accentuates the divide between the observer and the body in pain. Thus, the power of the image for the purpose of humanitarian communication is in the subject as referent object of the threat (such as bombing, shelling, famine or gas attacks) or rather as symbol of future similar threats to those already experienced that pain from which it could not be protected.

The problem is that, as in the case of representations of pity, "it is only in a world from which suffering has bene banished that justice could enforce its rights" (Boltanski 1999, 5). In fact, even images that have been published to arouse conscience and denounce what is shown at the visual level remains trapped in a visual short-circuit in which the body in pain is extremely visible, but the threat is not. Therefore the possibility of interventions is difficult to imagine outside of the protection of that body. None of the pictures collected for this study shows bombings, weapons, belligerent groups, sieges, deportations, or refoulement (forcible deportation of those seeking refuge).[17] The suffering body incorporates the threat. The need for intervention is imaginable therefore only limited to the protection of that body.

6.3.5   Innocence

A man is posing for the camera with a white dove in his hands is the subject of an OXFAM picture. The image portrays them in full light through a medium-close shot whose interactive meaning corresponds to a far personal relationship (Kress and Van Leeuwen 1996, 125). We are not so close as to understand his feelings, as in close-ups, but close enough to be able to relate at a personal, rather than at a social level. The middle-aged man is well-groomed and well-dressed in clean clothes. The bird he is proudly holding aloft has its wings spread open, as if poised for flight. The photographer has captured the exact moment the bird opens its wings while still remaining firmly held. It is no coincidence: the dove, its wings spread for flight, is clearly an instantly recognizable symbol of peace. The background, although in

---

[17] A core principle of international human rights and refugee law, the principle of *non-refoulement* guarantees that no one should be returned to a country where they would face torture, cruel, inhuman or degrading treatment or punishment and other irreparable harm.

sharp focus does not tell too much to a Western audience except that there is a roughly plastered wall and corrugated iron panels behind the man. To those who are more familiar with the urban landscape of some Middle Eastern cities, the setting may recall the typical visual landscape of rooftops where many people, generally men, practice a popular hobby in Syria, Lebanon and Jordan: homing pigeons.

The photo is emblematic of the new kind of humanitarian pictures that are meant to go beyond the traditional (pathetic) communication to show people not only with dignity but empowered, visual representation that tries to use positive images and tell 'success stories'. As an Oxfam advisor told me: "because the Syria crisis has evolved over time we are also trying to change images of refugees from the classical representation of refugees in tents (of course this element is still there), we are try- ing also to show empowerment and success stories with a particular focus on women and families. This is also because of the social situation that is showing more and more social tension and we try to produce a positive account."[18]

Indeed, this picture provides an alternative narrative of Syrian refugees. Not only does it portray a man, instead of the more typical focus on children, women and other representations of vulnerability, but it also depicts the person doing something other than just standing around. The caption unusually[19] indicates his name and surname: Jamal Ashamed. Dignity. Empowerment is suggested through the visual clue of the hobby. Pigeon fancying is something ordinary, part of Jamal's every day, and not confined to his displacement situation. This does change the narrative and certainly is a step forward from the more typical depiction of refugees as passive victims. The intention of representing the success story of a man empowered and with dignity is also confirmed by the text accompanying the image: "Jamal and his seven children fled Syria 3 years ago. He left everything behind but has managed to reclaim part of his identity, his passion for training pigeons. Jamal spent some time volunteering with Oxfam, before he was offered the opportunity to work with us as a water distributor." (OXFAM 2016).

While appreciating the effort behind this picture to provide a different account of Syrian displacement, its alternative and positive perspective, I would like to focus on the ideological significance of one of the image's metaphors. For the dove, a visual theme widely recurrent in pictorial art[20] besides being the ultimate symbol of peace and used in religious paintings to symbolize the holy spirit, also has another symbolic meaning, that of innocence.

From the biblical verse "I send you out as sheep in the midst of wolves. Therefore be wise as serpents and harmless as doves" (Matthew 10:16) to the poetic fantasies of William Blake's *Auguries of Innocence* "A Dove house filld with Doves & Pigeons Shudders Hell thr' all its regions" (Auden et al. 1950), the image of the dove – with its characteristic features of purity and immaculacy – is often used as

---

[18] Interview with an OXFAM Policy Advisor, Beirut, Lebanon, 8th March 2017

[19] As often stated in the text accompanying pictures, the names of the people portrayed in the pic- tures are usually changed to protect their identities.

[20] For example, Pablo Picasso's *Child with a Dove* (1901), Rene Magritte's *The Man with a Bowler Hat* (1964), or Marc Chagall's *Child with a Dove* (1937)

signifier of innocence. Despite the presence of the white bird the picture of Jamal Ashamed featuring on Oxfam website in 2016, makes it particularly emblematic, this is not the only image depicting refugees as innocents. On the contrary, this kind of representation is very common among NGOs pictures. I have already touched on the subject, discussing how the general over-representation of children and women, older people and those with disabilities in relief organizations' communications underlines the humanitarian subjects' vulnerability as it presents them as threatened victims.

Here, I would like to focus on the aspect of innocence. For this section, I decided to select images of men exactly because masculinity is a category usually not associated with innocence, especially when other groups of subjects are available for the camera. Obviously, the representation of innocence is easily visible and noticeable in the overwhelming presence of children and vulnerable groups in humanitarian images. By showing that it is also found in images of men, I hope I will be able to make my point even clearer.

As noted above, among all the images considered for this study neither in the case of images with a single represented participant, nor in those portraying groups, were men significantly less frequently photographed than women. Although dominant subjects of pictures of masses of people on the move, the large majority of the rest of the photos – across all four NGOs – depicted adult males with their families, their children, in caring and protective poses or in situations of vulnerability. The advantage of these kind of images is that they challenge stereotyped representation of refugees (particularly adult men) as threatening. They do so by focusing on their affections, education, or parental/protective role. All these images, added to the others that depict subjects as inherently innocent, clearly constitute a dominant visual narrative in NGOs' humanitarian communication. Again acknowledging the important effort of these different aesthetic patterns in depicting people on the move as threatened instead of threatening, there is something that I find problematic in the humanitarian representation that emphasize their innocence. In a "moral economy" in which discourses and practices around people on the move oscillate between compassion and repression (Fassin 2005, 365), there is little doubt that NGOs favor the former.

Generally, as Slim has pointed out, NGOs lean toward an "innocence-based solidarity" (Slim 1997, 350): a conceptualization of solidarity in which the principle is only claimed toward the 'innocents', often women and children, typically considered the most vulnerable, but also, as the pictures above show, innocent men. The problem is that this emphasis on innocence contributes also to the reification of dichotomous categories of innocents vs. non- innocents, deserving vs. undeserving refugees or recipients of humanitarian aid and non-beneficiaries

The importance of labeling a particular group of displaced people struck me forcibly when during my fieldwork in Lesbos. I visited Moria, Kara Tepe, and Pikpa refugee camps. The first was at the time the largest reception site on Greek islands and was set up by the government to process new arrivals and host refugees waiting for registration. Kara Tepe is a municipally-managed camp intended to accommodate families and vulnerable groups. Pikpa is a small reception site managed by a voluntary organization and hosting only vulnerable groups and especially unaccompanied and separated children. In Kara Tepe, I was initially puzzled by a remark from a camp manager that "we do not call them refugees, they are villagers.". I kept asking myself what could possibly be so bad about the term 'refugee' that it needed to be substituted by the artificial and inappropriate appellation/euphemism of 'villager.' Evidently, villagers needed to be protected from the stigma of being refugees as if this category intrinsically embodied something negative.

The difference between the concept of 'villagers' and 'refugees' lies in the action of moving, the decision to leave one place to go to another. It was as if refugees had lost innocence in this choice of moving, as if rights need to be threatened outside of any individual agency to be worthy of protection. In fact, as Friese has argued, despite agency and autonomy defining the modern subject, agency is threatening: "a helpless, needy and dependent person can be pardoned for that which in reverse marks the other as threat: namely autonomy, choice, and decision"' (Friese 2017, 6). My reflection was reinforced by the different demographics in each camp. While Moria at the time mostly housed single men – n – referred to as refugees – Kara Tepe was for families (including men) and vulnerable populations (Jauhiainen 2017), the innocents par excellence – referred to as 'villagers'. The denomination by the Kara Tepe camp management implicitly implied the perceived level risks associated with the two different populations.

In discussing seemingly antithetical security and humanitarian articulations concerning human trafficking, Aradau (2004) analysed how humanitarian organizations work with trafficked women. In order to move on from security approaches based on the politics of risk they prefer a form of representation capable of eliciting a response based on the politics of pity. Pity is, indeed, activated by the identification of the trafficked women as worthy of compassion as non-dangerous subjects. It must be noted that Aradau does not mean to suggest that these organizations aim to equate compassion with the supposed innocence of the victims. On the contrary, the notion of compassion is used by humanitarian organizations to go beyond the dialectic of victim-criminal according to which women in Aradau's case study are either women who have embarked consciously on an irregular journey or are innocent victims of criminal organizations. In this way the strategy of pity manages to erase any consideration on innocence and guiltiness.

The images of pity are functional in avoiding dichotomous representation of refugees as innocent vs. guilty – since they are all worthy of compassion. On the contrary, images of innocence, not based on the aesthetic pattern of pity, contribute to reproduce and reinforce a binary understanding of people on the move, one divided between those who are innocent and vulnerable *versus* the not innocent, those who do not deserve assistance. Along those lines, the diffusion of the visual pattern of innocence in the representation of people on the move has a similar tendency to challenge media representation that have consistently depicted people on the move as irregular migrants constituting a threat to state sovereignty. However, even acknowledging NGOs' efforts to question securitized accounts, the conceptual implication of diffusion of the visual pattern of innocence in the representation of people on the move implies that protection is for non-dangerous victims, the innocents. In this way, NGOs are leaning toward a defensive narrative in which in order to be considered worthy of protection refugees need to be shown to be innocent. Instead of being innocent until proven guilty, refugees fall into a perverse (and securitized) dynamic in which the burden of proving their innocence falls back on them.

6.3.6   The Savior Hero (Fig. 6.5)

Humanitarian narrative is often based on a quite predictable plot, what Chandler has called a "fairy story" in which: the (helpless) victim, the villain, the non-Western actor(s) responsible for violence, famine, and suffering and finally the savior, an external agent, typically humanitarian organizations, "whose interests were seen to be inseparable from those of the deserving victim" (Chandler 2002, 36). The visual narrative follows a very similar pattern but with a substantial difference: the villain is typically invisible.[21] While the previous sections have extensively explored the visual representation of the victim, I would like to focus now on the character of the savior. The image portrays a young Western woman holding a baby in her arms. Prominent in mid frame, the logo of the organizations she is working is emblazoned on her T-shirt. On the left of the frame, her back turned to us, there is an observer, a woman with veil or hijab, most probably the mother of the child. She is an external spectator and, more or less, like us. She is watching the protagonists of the image

Fig. 6.5 ©Anna Pantelia/Save the children

---

[21] See also Sect. 7.2.3 of the following chapter

while they are looking at each other. Whereas the humanitarian worker is smiling, the baby is looking bewildered, as children often are with strangers. We can deduce from the tents that the scene is set in an informal camp. For its semiotic characteristic (no direct eye contact between the represented participants and the viewer, medium/far shot, and frontal angle) this is clearly an 'offer' image, one meant to offer us some kind of information about the situation. What this image is telling us is that the organization is willing and able to assist the victims in their condition of displacement. The hug symbolically stands for the protection activities that constitute one of the core programs of many relief organizations.

Save the Children image is not unique in his genre. Several other pictures provide the exact same narrative (and pictures of this kind have been found to more present in MSF and Save the Children visual representation). In a thoughtful analysis of humanitarian visuality, the description that Kurasawa gives this visual theme could be easily applied to many other pictures of this kind: "for her or his part, the aid worker is demarcated from other actors in the image through situational and compositional symbols designating him or her as a benevolent, selfless and often courageous actor who intervenes in a humanitarian crisis to save the lives of victims, provide them with care, and/or ensure their recovery; hence, in addition to their corporeal poses and gestures (the examining or feeding of a subject, the carrying of supplies, etc.,) aid workers' relations to the crisis and its victims are visually signified through their clothing (e.g., a humanitarian NGO's t-shirt, a nurse's or doctor's uniform) and equipment (medical supplies, foodstuffs, aid tents, etc.)" (Kurasawa 2015, 17).

In the other images of this genre not only are organizational logos invariably visible and in center, but also the humanitarian workers are standing as if there were no other adults present, as if the present and future of the children they are maternally caring for solely depended on them. No mothers, fathers, or families are included in the pictures. Those who have nurtured, loved and probably saved their children' lives escaping violence are not represented in many of these pictures. With a few notable exceptions, when parents are captured by the camera, they are relegated to a corner or background, as if their role were marginal.

The visual pattern of the white hero assumes even more relevance if we think about what it tells us about humanitarian organizations' self-representation. The iconography of the savior resonates with the colonialist and paternalist approaches to the 'other' that I discussed in the section on *infantilization*. It is no coincidence that at the iconographical level these kinds of representation once again privilege children (and to a lesser extent women), and very rarely men. The pose of the girl in another image in which she walks toward the horizon hand-in-hand with two children with her chin held high and a spring in her step, suggests enlightenment and confidence in the action she is carrying out. For she is leading the children toward an invisible destination (metaphorically towards a brighter future?). The rhetoric of this image, whose hyperbolic representation risks sliding into irony, is not very far from images that have been criticized precisely for their arrogant and culturally imperialist approach. Parodic social media projects like Barbie Savior[22] or Humanitarians of Tinder[23] have highlighted how the "victim/hero" narrative reinforces paternalistic and colonialist approaches to humanitarian action and how the self-representation of the "white savior" can help us understand how power dynamics are reproduced.

Other than relief agencies self-representation, the visual theme of the victim-hero also reveals something about ourselves (spectators), specifically if we pay attention with whom we identify with in the pictures. Given the marginality or absence of adult migrants, we are inclined to identify with the other adults in the picture: the humanitarian workers. In an article discussing the double narrative (military and humanitarian) used by the Italian Navy in representing its engagement in the Operation Mare Nostrum,[24] Musarò (2017) shows how most of the images inspire sympathy for the armed forces personnel and pity for the rescued migrants. In emphasizing discourses of assistance and protection, the Navy represented its operation as actions of national benevolence, creating an "imagined community" (Anderson 1991) between the military personnel and the viewers from the same country: a community in which 'we', the spectators, identify with the saviors rescuing the 'others'. In the pictures discussed here a similar dynamic is at play except that the humanitarian goal is the NGOs' primary aim (and not an accessory one as in the case of the Italian Navy). The hero visual trope reinforces a particular narrative that constructs two imagined communities: the Western actors who altruistically and bravely intervene to rescue the helpless migrant. This kind of representation, as Musarò points out, "construct borders at imaginary levels, through mediated representations that are often presented as binary opposites" (Musarò 2017, 18).

In so doing, the aesthetic pattern of the savior/hero produces an account of humanitarian action that resonates much more with the idea of the unilateral action of an external agent who is in charge and knows what he is doing and a victim who is passively at the mercy of the events. As Kurasawa has underlined: "images of rescue are symbolically structured to visually inscribe the hierarchically organized roles and vastly unequal capacities of actors in the face of humanitarian crises, through the juxtaposition of positions between a rescuer possessing the agency to intervene to transform dire circumstances and a victim portrayed as helpless, passive or acutely vulnerable" (Kurasawa 2015, 25).

---

[22] Barbie Saviour is a social media project with a parody Instagram account (https://www.instagram.com/barbiesavior/?hl=es) created to express the founders' frustrations at stereotyped and victimizing representation of the Global South and what they term "the Barbie Savior complex "of many humanitarian volunteers. See: https://www.bbc.com/news/world-africa-36132482

[23] Humanitarians of Tinder (www.humanitariansoftinder.com) is a collection of images of people using their "humanitarian' encounter with the 'other" to define themselves. What started as a joke, gained traction and sparked media and scholarly discussion on how both professional and volunteer humanitarians represent themselves (Bex and Craps 2016; Richey 2016; Mason 2016).

[24] A year-long naval and air operation launched by Italy in October 2013. Its suspension is credited with helping to subsequently increase the numbers of refugees drowning in the Mediterranean.

This dichotomous and asymmetric representation, typical of the different visual registries depicting the threatened victims that we have analyzed in this chapter, inspires an idea of humanitarianism that is far from the declared core values of partnership, empowerment, and solidarity set out in the NGOs mandates and mission statements. Indeed, while aid is conceptually based on the idea of donation, and the logic of gift implies a hierarchical relation (Mauss 1925), solidarity is based on the idea of equality. At the visual level, equality is nowhere to be found, neither in the images of hero/savior or in images depicting NGOs staff and beneficiaries involved in humanitarian activities. This representation does indeed contribute to a securitized account of Syrian refugees, re-relegated to a victim role, passively waiting for the intervention of the humanitarian hero.

6.4 Conclusion

This chapter has addressed the different visual patterns through which NGOs represent Syrian people on the move as referent object of a threat. Although the topic of the threatened individual is depicted in quite diverse aesthetic themes, they all describe displaced people at risk and in need of humanitarian protection. At the visual level, the combination of all the semiotic resources that contribute to the representation of the victims as threatened are part of the 'new' attention to human security and the concomitant shift from a needs-based to a rights-based approach to humanitarian intervention.

I have shown how images of pity, with their similarity with the religious iconography of piety, have the power to inspire an emotional response that uses the language of urgency instead of that of justice. In this way, the only imaginable action is that of immediate intervention to interrupt the suffering, while the wider political context and the causes of the distress are completely overlooked. Moreover, this kind of visual rhetoric has shown an over-representation of women that is at odds with the general trend for humanitarian communication images to depict men and women in equal measure. This reinforces the construction of the female body as intrinsically vulnerable.

The exploration of NGOs photographic material has also confirmed an account of people on the move as essentialized victims, a phenomenon extensively discussed in the literature on refugees and humanitarianism. The depiction in terms of helplessness and passiveness is underlined by the over-representation of children that I observed consistently in the picture studies for this research. The overwhelming presence of minors as photographic subjects contributes to the reproduction of an infantilized, colonial, and paternalistic account of people on the move.

The visual analysis of humanitarian organizations' photos has also shown how very vivid images of suffering and physical pain, produced with the evident intention to draw attention to critical situations, have also the opposite effect of distancing the viewer from the represented participant, underlying the asymmetry of position between them. Moreover, the body in pain comes to incorporate and represent the threat, while at the same time concealing it. What remains is the sense of urgency for a protective intervention that once again speaks in terms of emergency instead of justice.

Throughout the analysis of the symbolic value attributed to many images of Syrian people on the move, I have also shown how a special emphasis is placed on innocence. Although undoubtedly intended to challenge media and political accounts of refugees as threatening, this kind of depiction has also the effect of reifying dichotomous categories: innocents/non-innocents, deserving/undeserving or recipients/non recipients of humanitarian aid. The last visual trope that I have explored is the 'hero', a type of humanitarian self-representation through which NGOs present their essential selflessness. This dichotomous and asymmetric visual pattern does more than resonate with the colonial and paternalistic approach underlined in the case of infantilization. For it also contributes to the creation of binary 'imagined communities' of universal heroes and universal victims, an involuntary irony, satirical representations of present-day humanitarianism.

It is important to notice that the investigation of humanitarian organizations' visual production has also highlighted a new kind of representation that aims to go beyond a portrayal of beneficiaries in terms of victimhood. As I have shown in the previous sections, relief organization have put a great effort in changing their representation of the beneficiaries, including victims of displacement. Clearly, one of their core concerns has been to stop portraying them as victims, opting for a depiction of beneficiaries showing their dignity and empowerment. NGOs' intentions in this sense are confirmed by their communication guidelines (Manzo 2008; Lidchi 1999) and the interviews I conducted during fieldwork.

These 'new' type of images based on determined visages and confident faces, while putting the emphasis on people's rights – not only to survival but also integrity, dignity, and wellbeing – also contributes to the shaping of the protection authority over the person of concern. In this political act, the individual whose rights need to be protected is securitized by the humanitarian intervention that become the most legitimate security strategy to address the threat. Indeed, although the focus on the threatened rights seems to challenge securitized discourses on migration based on security and control, these two perspectives are not an alternative framing. On the contrary, "securitization and securing rights are, in fact, two sides of the same coin" (Sasse 2005, 689).

For these reasons, all these different depictions of Syrian people on the move as threatened have failed to fulfill the emancipatory promise of the rights-based human security framework. Indeed, the diverse visual themes, far from empowering or emancipating the humanitarian subjects, weaken any political claim they may have, and rather inscribe them into a dynamic of dependency requiring external protection. Against this background, humanitarian protection has increasingly gained momentum as the operationalization of a rights-based approach aimed at safeguarding the dignity, integrity, and wellbeing of the individual, and no just immediate life-saving assistance. However, its implementation is strictly interrelated with securitization discourses and practices as protection has become one of the most legitimate security strategies to address human insecurity.

References

Achilli, L., Hannah, L., Monica, M., Marina, T., Alba, C., & Triandafilia, K. (2017). *On my own: Protection challenges for unaccompanied and separated children in Jordan, Lebanon and Greece*. Technical report, Mixed Migration Platform (MMP) Report, INTERSOS, Migration Policy Centre.

Anderson, B. (1991). *Imagined communities: Reflections on the emergence of nationalism*. London: Verso.

Aoláin, F. N. (2011). Women, vulnerability, and humanitarian emergencies. *The Michigan Journal of Gender & Law, 18*, 1.

Aradau, C. (2004). The perverse politics of four-letter words: Risk and pity in the securitisation of human trafficking. *Millennium, 33*(2), 251–277.

Arendt, H. (1990). *On revolution*. London: Penguin Books.

Auden, W. H., Pearson, N. H., & Talbot Donaldson, E. (1950). *Poets of the English language* (Vol. 4). New York: Viking Press.

Bex, S., & Craps, S. (2016). Humanitarianism, testimony, and the white savior industrial complex: What is the what versus Kony 2012. *Cultural Critique, 92*(1), 32–56.

Bleiker, R., Campbell, D., & Hutchison, E. (2014). Visual cultures of inhospitality. *Peace Review, 26*(2), 192–200.

Boltanski, L. (1999). *Distant suffering: Morality, media and politics*. Cambridge: Cambridge University Press.

Bourgois, P. (2001). The power of violence in war and peace: Post-Cold War lessons from El Salvador. *Ethnography, 2*(1), 5–34.

Brems, E. (1997). Enemies or allies? Feminism and cultural relativism as dissident voices in human rights discourse. *Human Rights Quarterly, 19*(1), 136–164.

Burman, E. (1994). Poor children: Charity appeals and ideologies of childhood. *Changes-Sheffield, 12*, 29–29.

Butler, J. (2004). *Precarious life: The powers of violence and mourning*. London: Verso.

Byrne, B., & Baden, S. (1995). *Gender, emergencies and humanitarian assistance*. Report commissioned by the WID desk, European Commission, Directorate General for Development 33. BRIDGE (Development – Gender) Institute of Development Studies.

Calain, P. (2012). Ethics and images of suffering bodies in humanitarian medicine. *Social Science & Medicine, 98*, 278–285.

CARE. (2016a). *CARE Syria response – Posts*. https://www.facebook.com/CARESyriaResponse/photos/a.362149093938941.1073741828.361464940674023/661852677301913/?type=3&theater

CARE. (2016b). *Donate*. Text. CARE. http://www.care.org/donate

Care International. (2017). *Care International annual report FY16*. https://www.care-international.org/files/files/publications/Care_International_Annual_Report_2017_8921_ENG_12_Online(4).pdf

Carling, J., & Hernández-Carretero, M. (2011). Protecting Europe and protecting migrants? Strategies for managing unauthorised migration from Africa. *The British Journal of Politics and International Relations, 13*(1), 42–58.

Chandler, D. (2001). The road to military humanitarianism: How the human rights NGOs shaped a new humanitarian agenda. *Human Rights Quarterly, 23*(3), 678–700.

Chandler, D. (2002). *From Kosovo to Kabul: Human rights and international intervention*. London: Pluto Press.

Chouliaraki, L. (2010). Post-humanitarianism: Humanitarian communication beyond a politics of pity. *International Journal of Cultural Studies, 13*(2), 107–126.

Chouliaraki, L. (2013). *The ironic spectator: Solidarity in the age of post-humanitarianism*. John Wiley & Sons: Cambridge.

Citton, Y. (2014). *L'économie de l'attention: Nouvel Horizon Du Capitalisme?* La Découverte.

Concord. (2006). *Code of conduct on images and messages*. https://concordeurope.org/wp-content/uploads/2012/09/DEEEP-BOOK-2014-113.pdf?c676e3&c676e3

Coomaraswamy, R. (1994). To bellow like a cow: Women, ethnicity, and the discourse of rights. *Human Rights of Women: National and International Perspectives, 39*, 56.

Dauphinée, E. (2007). The politics of the body in pain: Reading the ethics of imagery. *Security Dialogue, 38*(2), 139–155.

DuBois, M. (2007). Protection: The new humanitarian fig-leaf. *Dialogue, 4*, 1.

Duffield, M. (2014). *Global governance and the new wars: The merging of development and security*. London: Zed Books Ltd.

Duffield, M., & Waddell, N. (2006). Securing humans in a dangerous world. *International Politics, 43*(1), 1–23.

Encyclopaedia Britannica. (2018). *Metonymy*. Figure of speech. Britannica.Com. https://www.britannica.com/art/metonymy

Fassin, D. (2005). Compassion and repression: The moral economy of immigration policies in France. *Cultural Anthropology, 20*(3), 362–387.

Fassin, D. (2007). Humanitarianism as a politics of life. *Public Culture, 19*(3), 499–520.

Fehrenbach, H., & Rodogno, D. (2015). "A horrific photo of a drowned Syrian child": Humanitarian photography and NGO media strategies in historical perspective. *International Review of the Red Cross, 97*(900), 1121–1155.

Fox, F. (2001). New humanitarianism: Does it provide a moral banner for the 21st century? *Disasters, 25*(4), 275–289.

Friese, H. (2017). Representations of gendered mobility and the tragic border regime in the Mediterranean. *Journal of Balkan and Near Eastern Studies, 19*, 1–16.

Heck, A., & Schlag, G. (2013). Securitizing images: The female body and the war in Afghanistan. *European Journal of International Relations, 19*(4), 891–913.

Holland, P. (1992). *What is a child? Popular images of childhood*. London: Virago.

Human Rights Watch. (2016a). Growing up without an education. Barriers to education for Syrian refugee children in Lebanon. *Human Rights Watch*, 19 July. https://www.hrw.org/report/2016/07/19/growing-without-education/barriers-education-syrian-refugee-children-lebanon.

Human Rights Watch. (2016b). *We're afraid for their future. Barriers to education for Syrian refugee children in Jordan*. https://www.hrw.org/report/2016/08/16/were-afraid-their-future/barriers-education-syrian-refugee-children-jordan

IASC. (2006). *Gender handbook in humanitarian action*. Inter-Agency Standing Committee.

Jauhiainen, J. (2017). Asylum Seekers in Lesvos, Greece, 2016–2017. *Turun Yliopiston Maantieteen Ja Geologian Laitoksen Julkaisuja, 6.*

Kennedy, D. (2009). Selling the distant other: Humanitarianism and imagery – Ethical dilemmas of humanitarian action. *The Journal of Humanitarian Assistance, 28*, 1–25.

Kleinman, A., & Kleinman, J. (1996). The appeal of experience; the dismay of images: Cultural appropriations of suffering in our times. *Daedalus, 125*, 1–23.

Kress, G. R., & Van Leeuwen, T. (1996). *Reading images: The grammar of visual design*. London: Psychology Press.

Kuperman, A. J. (2013). A model humanitarian intervention? Reassessing NATO's Libya campaign. *International Security, 38*(1), 105–136.

Kurasawa, F. (2015). How does humanitarian visuality work? A conceptual toolkit for a sociology of iconic suffering. *Sociologica, 9*(1). https://doi.org/10.2383/80396.

Lidchi, H. (1999). Finding the right image: British development NGOs and the regulation of imagery. In *Culture and global change* (pp. 87–101). Florence: Taylor and Francis.

Malkki, L. H. (1995). Refugees and exile: From "refugee studies" to the national order of things. *Annual Review of Anthropology, 24*(1), 495–523.

Malkki, L. H. (1996). Speechless emissaries: Refugees, humanitarianism, and dehistoricization. *Cultural Anthropology, 11*(3), 377–404.

Manzo, K. (2008). Imaging humanitarianism: NGO identity and the iconography of childhood. *Antipode, 40*(4), 632–657.

Martin, M. (1985). *Le Langage Cinématographique.* Paris: Editions du Cerf.

Mason, C. L. (2016). Tinder and humanitarian hook-ups: The erotics of social media racism. *Feminist Media Studies, 16*(5), 822–837.

Mauss, M. (1925). *The gift: The form and functions of exchange in archaic societies.* London: Routledge.

Moeller, S. D. (1999). *Compassion fatigue: How the media sell disease, famine, war and death.* London: Routledge/Psychology Press.

Musarò, P. (2013). "Africans" vs. "Europeans": Humanitarian narratives and the moral geography of the world. *Sociologia Della Comunicazione, 45*, 37–59.

Musarò, P. (2017). Mare nostrum: The visual politics of a military-humanitarian operation in the Mediterranean Sea. *Media, Culture and Society, 39*(1), 11–28.

Ogata, S., & Sen, A. (2003). *Final report of the commission on human security.* United Nations Commission on Human Security.

Oxfam. (2013). *The power of people against poverty. Oxfam strategic plan, 2013–2019.* https://www.oxfam.org/sites/www.oxfam.org/files/oxfam-strategic-plan-2013-2019.pdf

OXFAM. (2016). *Life in Za'atari Refugee Camp, Jordan's Fourth Biggest City.* Oxfam International. https://www.oxfam.org/en/crisis-syria/life-zaatari-refugee-camp-jordans-fourth-biggest-city

Pupavac, V. (2001a). Misanthropy without Borders: The international children's rights regime. *Disasters, 25*(2), 95–112.

Pupavac, V. (2001b). Therapeutic governance: Psycho-social intervention and trauma risk management. *Disasters, 25*(4), 358–372.

Pupavac, V. (2005). Human security and the rise of global therapeutic governance. *Conflict, Security and Development, 5*(2), 161–181. https://doi.org/10.1080/14678800500170076.

Rajaram, P. K. (2002). Humanitarianism and representations of the refugee. *Journal of Refugee Studies, 15*(3), 247–264. https://doi.org/10.1093/jrs/15.3.247.

Richey, L. A. (2016). "Tinder humanitarians": The moral panic around representations of old relationships in new media. *Javnost-The Public, 23*(4), 398–414.

Rieff, D. (2003). *A bed for the night: Humanitarianism in crisis.* New York: Simon and Schuster.

Rosenberg, E. S. (2002). Rescuing women and children. *The Journal of American History, 89*(2), 456–465.

Sasse, G. (2005). Securitization or securing rights? Exploring the conceptual foundations of policies towards minorities and migrants in Europe. *JCMS: Journal of Common Market Studies, 43*(4), 673–693.

Save the Children. (2016a). Children on the move in Europe by Save the Children. *Exposure.* https://savethechildreninternational.exposure.co/children-on-the-move-in-europe

Save the Children. (2016b). *Save the children's global strategy: Ambition for children 2030 and 2016–2018 strategic plan.* https://www.savethechildren.net/sites/default/files/Global%20Strategy%20-%20Ambition%20for%20Children%202030.pdf

Schubert, V. (2007). 'Strange bedfellows?' Security, humanitarianism and the politics of protection. *Melbourne Journal of Politics, 32*, 25.

Slim, H. (1997). Relief agencies and moral standing in war: Principles of humanity, neutrality, impartiality and Solidarity1. *Development in Practice, 7*(4), 342–352.

Slim, H. (2003). Is humanitarianism being politicised? A reply to David Rieff. In *The Dutch Red Cross Symposium on Ethics in Aid,* The Hague, 8th October 2003, Vol. 8.

Slim, H., & Bonwick, A. (2006). *Protection: An ALNAP guide for humanitarian agencies.* Warwickshire: Oxfam.

Sontag, S. (1973). *On photography* [Online]. Electronic Edition 2005. New York: RosettaBooks LLC.

Sontag, S. (2003). *Regarding the pain of others.* London: Hamish Hamilton.

The Independent. (2015). This is how Europe is welcoming refugees right now. *The Independent*. http://www.independent.co.uk/news/world/europe/hungary-opens-fire-on-refugees-with-tear-gas-as-it-steps-up-border-operation-10503822.html

Time. (2015). Are the Syrian refugees all "young, strong men"? *Time*. http://time.com/4122186/syrian-refugees-donald-trump-young-men/

UNHCR. (2016). *Situation Syria regional refugee response*. https://web.archive.org/web/20160919092358; https://data2.unhcr.org/en/situations/syria

UNICEF/REACH. (2014). *Syrian refugees staying in informal tented settlements in Jordan: Multi-sector needs assessment*. https://reliefweb.int/sites/reliefweb.int/files/resources/REACH_UNICEF_ITS_MS_AUGUST2014_FINAL.PDF

Van Leeuwen, T., & Jewitt, C. (2001). *The handbook of visual analysis*. Los Angeles: Sage.

Watson, S. (2011). The 'human' as referent object? Humanitarianism as securitization. *Security Dialogue, 42*(1), 3–20.

Žižek, S. (2005). Against human rights. In *Wronging rights?: Philosophical challenges to human rights* (pp. 149–167). New York: Routledge.

Chapter 7
The (In)Visibility of Migrants

7.1 Introduction

Bruno Catalano's sculpture *Voyageurs*[1] brilliantly catches an impalpable, yet perva-sive, feature of refugees' representation: their invisibility. This may seem an oxy-moron but throughout my visual analysis, and while looking for what was represented in the images studied, I have been struck by what is *not* there. Susan Sontag has defined photography as "grammar and, even more importantly, an ethics of seeing!" (Sontag 1973, 1). If Sontag is right when she affirms that photography makes things represented worth being seen, we may be led to think that, on the contrary, what is not photographed is not worth seeing. On this arbitrary decision over presence and absence, visibility and invisibility, hinges a very important dimension of the power of photography.

In this sense, representation is as much about presence as about absence (Manzo 2008). Visual studies have recognized the importance of the unseen (Foster 1988) and the need to conceptualize the absence (Rogoff 2000). A crucial aspect of (in) visibility is its role in social, political and security practices (Ericson and Haggerty 2006). Rancière (2014) has pointed out how systems of meaning function at the visual level, limiting or encouraging thought. Similarly, Bleiker et al. (2013) have argued that by their way of shaping what can (and cannot) be seen, images indi-rectly shape what can (and cannot) bethought. Even more, images affect the possi-bility of political positions being adopted, or opinions expressed publicly (Bleiker et al. 2013). Paying attention to Rancière's system of visibility – "aesthetic regimes" that decide what should be and should not be visible – allows us to reach a better understanding of refugees' representation and the relevance of what is left out of the picture.

---

[1] Images of Bruno Catalano's sculptures, including the voyageurs series, can be seen at: www. brunocatalano.com

© The Author(s) 2021
A. Massari, *Visual Securitization*, IMISCOE Research Series,
https://doi.org/10.1007/978-3-030-71143-6_7

Invisibility is not only relevant *per se,* but also for its dialectic relation with visibility. It is by paying attention to what is recurrently present that enables us to identify the significant omissions in how NGOs represent Syrians on the move. In a groundbreaking visual analysis of media coverage of the Darfur conflict, Campbell (2007) observes how the photographic narrative has been consistently built around the humanitarian dimension, while the theme of genocide – that initially inspired journalistic interest and remained an important topic at the textual level – remained systematically absent from the visual representation. In a similar way, exploring what is made visible and what is not in the humanitarian photographic account of the Syrian refugee 'crisis', allows us to shed light on how the dialectic between visible and invisible plays out in the representation of Syrian displacement.

(In)visibility is also a particularly interesting lens through which to explore images to highlight their potentiality in terms of emancipation and oppression. On the one side – and as I have already discussed in the previous chapter – there is a wide consensus that invisibility does little good for the victims of violence (Sontag 2003; Kleinman and Kleinman 1996; Butler 2004). Historically, there are cases in which images have brought to public attention dramatic events that were mostly unknown to the general audience. The visual representation of the Biafran famine, along with Western media attention and mobilization that sprang from it, is probably one of the most famous cases in the history of humanitarian photography. From a psychological perspective as well, invisibility can have serious negative implications. Inspired by the publication of a novel on the African American striving for self-affirmation in the 1950s, Franklin (1999) introduced the notion of "Invisibility Syndrome." This is psychological experience whereby people feel that their identity is consistently undermined by racism and they have to constantly struggle to obtain value and recognition. In these cases, it is quite evident how emancipation passes (also) through visibility. Visual representation of dramatic events of violence or injustice can be ethically very questionable. The picture of the starving Sudanese baby with the vulture that won the Pulitzer prize in 1994 (Carter 1993) is a paradigmatic example of this kind of photo. Kleinman and Kleinman (1996) observed how the long time spent by the photographer waiting for the vulture to fit the frame, instead of doing something to help the child in clear distress, the complete lack of context, and the mysterious absence of the parents suggest how little room for emancipation that image has produced. From a different perspective, the oppressive role of invisibility, and the asymmetrical power relation between the observer and the observed has been masterfully underlined by Foucault's discussion of the panopticon: a system of surveillance and governance ensured by the total visibility of the dominated subject and the invisibility of the controller (Foucault 1975).

In the specific context of the so-called migration crisis, while boats crowded with people have dominated the media account, the large number of migrants who have died (and continue to die) during the journey has remained mostly invisible (Falk 2010). With the important exception of the dramatically famous photo of Aylan Kurdi, there has been little emphasis on the thousands of migrants who have died along the route.[2] In Calais, French authorities have adopted a "'policy of invisibili-sation", removing migrants from public view to project an image that the migrant 'issue' has been solved." (Slingenberg and Bonneau 2017). These examples power-fully show how *absence*, and not just *presence*, requires attention in a visual analy-sis of how humanitarians account for refugees. With this, I am certainly not arguing that a higher level of visibility of suffering or death would have improved in any way the situation of Syrians on the move. However, an investigation of the visual presence and absence can improve our understanding of the dynamics of NGOs' representation of Syrian displacement and provide a sense of everything that is left out of n the picture.

From a critical security studies perspective, invisibility is also interestingly con-nected with the ability of visuality to contribute to securitization discourses. Since visibility works powerfully in the construction of both utterances (Campbell and Shapiro 2007), I argue that so does invisibility. The present chapter addresses four visual themes that are particularly paradigmatic in a discussion about presence and absence in the context of Syrian displacement. It first looks at politics, the big 'ele-phant in the room' of humanitarian discourse, to explore the extent to which politics is reflected in NGOs' visual materials. Secondly, the chapter explores the (in)visi-bility of refugees' protest and contestation in the difficult conditions of protracted displacement. The third section focuses on the (in)visibility of the threats to Syrians on the move, both inside Syria and in the countries of asylum. Finally, the last part investigates the absence of refugees' voices from humanitarian communication – despite NGOs' explicit strategic object to 'make people's voices heard.'

7.2   The Impalpable Importance of Invisibility

7.2.1   The Invisibility of Politics

When I first saw on one of the NGOs' websites a picture of a girl smiling and mak-ing the V for victory sign it made me immediately recall a couple of episodes that occurred during my visits to NGO-run community centers[3] in Jordan. In one case, I went to a child-friendly space – places increasingly used in emergency settings to promote child psycho-social wellbeing and strengthen local communities' capaci-ties in child protection – in Irbid. I was accompanied by the staff of a local NGO and a photographer who was documenting the organization's work in the field. When we entered the playground, interrupting the activities, a bunch of children ran toward us and a small group started posing for the camera. They were smiling and a couple

---

[2] For more information on missing migrants, see the International Organization for Migration (IOM) Missing Migrants Project, available at www.missingmigrant.iom.int.

[3] Community centers are facilities set up in humanitarian settings to provide safe spaces to dis-placed communities and where activities of various forms (from legal assistance to psychosocial support, from awareness-raising sessions to catch-up classes) are organized by NGOs.

were making the V sign while the photographer kept shooting pictures of this group of joyful children. The NGO's team coordinator told the photographer not to use those photos, explaining that for security reasons the organization did not want pictures of children making the V sign as the gesture could be linked to a specific political or conflict-related event in Syria, leaving the children or families exposed, both in Jordan and back home in Syria, for their apparent affiliation with a protagonist to the Syrian conflict. When visiting the field sites of other NGOs in Zaarqa, Maan, and Karak, I was struck by the quantity of Syrian flags – the pre-Baathist version with three red stars adopted by opponents of the Assad regime – in the children's drawings, affixed to walls and commonly painted on cheeks during face-painting activities. Despite its ubiquity I did not see a single picture containing this flag in the more than a thousand images collected for this study. The invisibility of the national symbol in NGOs' official photos was motivated by the same security concerns expressed in relation to the V sign.

Though understanding this reluctance, I, nevertheless, find the general absence of political signs and of politics from NGOs' visual material to be problematic. As Anna Leander has argued, "visibility is most political, when its politics shows least" (Leander 2016, 3). The invisibility of politics in the NGO representation of Syrian displacement is at odds with the importance that the civil conflict has had on people's lives. However, the invisibility of politics is not at odds with the idea that NGOs have of their work and its relationship with politics. When I probed NGO informants on the connections between humanitarianism and politics and how NGOs managed to accommodate the humanitarian principles of neutrality and impartiality in such highly political contexts, answers were quite clear. One communication officer declared: "our organization does not use 'political words'. For instance, we do not use the words Daesh, ISIS …we would rather say 'militant groups'. If a refugee says: 'my government is killing me' we will not reproduce it. We cannot take any political stance, it is a statement too strong, too politicized. For example, if I interview a refugee and he talks about ISIS I will have to reduce or change some words and make sure that there is not a strong political statement."[4] Another communication and advocacy specialist told me that "everything concerning refugees in Lebanon is about politics. It is no secret that the Syrian army was in Lebanon from 1976 to 2005 and not everybody appreciated this. Politics goes into that whether we like it or not. But we take a human rights perspective, we do not talk about politics. This position is the same with LHIF [Lebanon Humanitarian INGO Forum], Save the Children, Oxfam."[5]

These testimonies confirm that for NGOs avoidance of political statements is important to preserve their neutrality and impartiality. Political signs, working as visual symbols of political declaration, hinder the perception public opinion has of the organizations as not involved in politics but only interested in the pursuit of their humanitarian goals. I have already explored the debate on the complicated relations

---

[4] Skype interview with a CARE Emergency and Communication Officer, 13th June 2017.

[5] Interview with CARE Communication and Advocacy Manager, Beirut, 7th March 2017.

that humanitarianism has with politics in Chap. 3. What is relevant here is that politics, symbols and statements, although present both at the visual and discursive level among refugees, are intentionally absent in NGOs' communication material on the Syrian displacement, lest any dabbling in politics undermine the humanitarian message and intervention.

There is a further fundamental aspect of this invisibility. The absence of political signs in NGOs' visual representation makes even more sense if considered as part of their intention to challenge media and public accounts of refugees as a threat. Against the complex background of the Syrian civil war and the myriad of actors involved, a depiction of people on the move as completely detached from political affiliations, sympathies, recruitment or support is a way of defying securitized discourses on Syrian refugees. A field-based communication officer told me that for this exact reason they would not photograph armed people close to the hospital where they were working, even though many people arrived armed and the hospital even had a storage area for weapons.[6] More generally, humanitarian workers, and advocacy staff particularly, were very well aware of the role that their use of politics, particularly when linked with national security concerns in the host countries, could have on the public perception of refugees. As an advocacy specialist for Lebanon Humanitarian INGO Forum (LHIF) affirmed: "some of our members focus a lot on security: for example, on the campaign on education in emergencies. Their narrative comes from the connection to radicalization. The LHIF would not use the reference to the risk of 'recruitment by armed forces' as an argument in Lebanon because in Lebanon there is a very strong narrative on the instability of the situation and the military force of the different groups. Such an argument would work against the interests of the Syrians because there is the fear that the government will eventually kick out the Syrian refugees for security reasons. That is why the LHIF would never frame an issue in terms of radicalization, terrorism, and so on. Moreover, we do not want to contribute to the global narrative on security and securitization."[7]

At the same time, the absence of politics from NGOs' refugee representation also contributes to a dehistoricized and depoliticized depiction. In a seminal article discussing mass displacement and humanitarian dehistoricization, Malkki (1996) explains how for Hutu refugees talking about their displacement was talking about politics, the politics of the conflict that had led them to seek refuge abroad. But political discussions and political activism could not be encouraged by camp management as they were seen as incompatible with refugee status (Malkki 1996). This dehistoricizing process, she argues, is problematic in so far as it contributes to the creation of the category of universal victim. In the photographic account of Syrian refugees, something similar happens. Although as I have shown, it is a conscious and well-intended move inscribed in NGOs' protection and counter-securitization

---

[6] Interview with MSF Communication Officer, Amman, 18th May 2017.
[7] Interview with LHIF Advocacy & Communications Coordinator, Beirut, 15th March 2017.

strategies, the absence of politics contributes to a dehistoricized and depoliticized depiction of displaced Syrians.

7.2.2   The Invisibility of Protest

Connected with the absence of political signs, conflict-related symbols and political manifestations is the total invisibility of protest and mobilization among refugees in the way that NGOs communicate. Over the years, and given the often extremely harsh conditions of displacement, Syrians on the move have organized protests, some of which have involved radical acts such as hunger strike and sewing lips together, that have caught the attention of international media (The Independent 2015; Al Jazeera 2015; The Nation 2016). Some of these demonstrations have also had a strong visual expressiveness. More generally, different kind of refugees' mobilizations occurred within camps and informal settlements in 2015–2016 (Clarke 2017).

However none of this activism or any form of protest, sit-in, gathering, stone throwing or hunger strike is reflected in NGOs' visual representation. Similarly, there is no visual indication of any kind of individual frustration, resentment or anger that people on the move may have expressed regarding their condition of displacement, hospitality, access to third countries, or humanitarian assistance. This absence of discontent in depiction of refugees' lives is in line with some NGOs' communication objectives, not so much to provide information on refugees as to mobilize support for them.

On one hand, the fundamental humanitarian principle of 'do no harm' requires staff to be careful not to create further distress through their intervention. Emergency actors are thus very cautious not to fuel social instability and determined to foster social cohesion between different vulnerable communities. As one communications specialist told me: "Our general message is do no harm…of course it is a very slippery slope. For example, recently we have focused a lot on 'the neighbourhood approach'. It refers to the fact that hosting communities in Lebanon (including Palestinians) are very often in need as much as the refugee communities. Our approach is to work by sector of intervention and not according to specific target communities."[8]

Against this background, depicting protests organized by the recipients of humanitarian assistance risks depicting them as ungrateful or fuelling the grievances of communities aggrieved 1 their needs have not been (equally) addressed. One NGO media specialist explained that her team usually 'prepared' her agency's beneficiaries prior to a communication visit. She frankly told me that alongside describing the purpose of the meeting and asking for refugees' informed consent for the publication of statements, they would try to control the message. When I asked

---

[8] Interview with CARE Communication and Advocacy Manager, Beirut, 7th March 2017.

her to elaborate she told me: "if a journalist asks a question on livelihoods that are generating any income.[9] In that case, if the family is going to curse the government or the UN, we will change family. We want them to speak their minds, but we do not want them to fight with the government. What is really important is protecting the hosting communities. Also, for instance, if someone refers to the social tension problem, we do not want that in writing because it can offend someone, given the generosity of Lebanese society."[10]

The invisibility of protest resonates with the intention of humanitarian NGOs to challenge media and public accounts that present Syrians on the move as a threat. Showing demonstrations, acts of anger or dissent would contribute to reinforcing a securitized media and public discourse that has consistently depicted refugees as a threat. This strategy is in line with two desecuritizing options conceptualized by Wæver: avoiding reproducing a discourse about the issue in security terms and, given that the issue has been already securitized "to keep the responses in forms that do not generate security dilemmas and other vicious spirals" (Wæver 2000, 253). At the same time, the invisibility of protests in NGOs' visual communication precludes the option of what Huysmas (1995) has defined as a deconstructivist desecuritization strategy: to relocate the issue from the security dimension into the ordinary realm of politics. In this way, the absence in humanitarian depictions of Syrian refugees' dissent, demonstrations or political mobilization simultaneously contributes to produce multiple effects. These accord with humanitarian organizations' intention to not fuel a representation of people on the move as threatening, while also precluding their opportunity for 'making people voices heard.' The ultimate effect is to represent them as passive victims with limited agency.

An easy, yet erroneous, temptation would be to conclude that the invisibility of politics and protests is implemented as part of a strategy aimed at the protection of the humanitarian subject from securitized accounts, while actually serving the opposite role of silencing their political agency and depoliticizing their cause. What I would like to suggest here is that invisibility produces both meanings at the same time. As Yurchak (2013) has shown, it is neither a question of "official" versus "hidden" discourse nor the enactment of Marxian "false consciousness." NGOs' communication is intended for the protection of their beneficiaries from other (more or less vulnerable) communities and negative perceptions on the part of host governments which could put them at risk. While discouraging representation of the values, attitudes and actions of mobile Syrians that would fuel a hostile reaction, they at the same time, conceal any form of the refugees' political agency. From this eloquent invisibility emerges a passive and depoliticized subject. If the

---

[9] By way of explanation, it should be noted that the topic of livelihoods for Syrian refugees has been controversial both in Lebanon and Jordan since the beginning of the refugee crisis. The two governments, following a trend apparent in other refugee hosting contexts across the world, prevent relief organizations from implementing any kind of income-generating activities and vocational training on the ground that to do so would negatively affect local unemployment rates and fuel social discontent and anti-refugee sentiments.

[10] Interview with SAVE Media Advocacy and Communications Manager, Beirut, 8th March 2017.

depoliticization of people on the move seems self-evident when they are represented as threat in a securitized discourse that disregards any individuality, personal trajectory, or political context, in this case this happens precisely as a result of NGO discourses designed to counter such an overly securitized account.

7.2.3   The Invisibility of Threat (Fig. 7.1)

7.2.3.1   Bombing, Weapons, Tanks, and Soldiers

The figure is part of a larger group of pictures showing the destruction caused by the war. It is the typical picture that visual social semiotics defines as an information image: one that is not establishing any direct contact between the represented participants and the viewer, but where the represented participants – people standing in the middle of buildings in ruins – are rather 'offered' to the viewer as "items of information, objects of contemplation, impersonally, as though they were specimens in a display case" (Kress and Van Leeuwen 1996, 119). This does not absolutely imply that these kinds of images cannot elicit an emotional reaction. However, when they do so, it is a mechanism based on different premises than in the case in which the represented participants and the viewer are directly connected through the subject' gaze. The difference lies in the psychological implications of the eye contact between two human beings and the more detached representation with no

Fig. 7.1 ©CARE, Syria Response 19 December 2016

personal interrelation. In Fig. 7.1, the far distance of the shot and the deep depth of field allows us to observe the extent to which the war in Syria has caused destruction and led to displacement. The level of devastation is underlined by several visual elements: the smoke rising from the ruins – that leaves us wondering whether it is an effect of the recent bombing or something that has been burned to keep people warm – the confusion about the crowd's direction (since people seem randomly standing or going back to what would seem to be the direction of the convoy), the impossibility of seeing where everybody is trying to go since the picture does not allow for any hope. The sense of disorientation and uncertainty faced by the Syrians represented in the image is also confirmed by the post accompanying the picture: "For the past weeks the people of #Aleppo have suffered so much from the uncertainty of the situation. This must end" (CARE Syria Response 2016).

Despite the power of this image in showing the effect of war, what it is strikingly absent from this shot and all the others sharing the same visual themes of destruction, is the real threat, the perpetrator, the cause of such devastation. In many pictures, we can see the level of destruction and the physical suffering caused by violence and war, but we do not get to see the soldiers, the militia members, the bombs, the tanks, the warplanes that have been the causes of such destruction and suffering. The various kind of images that I consider here include photos with no human element or with people so tucked away in the far background that they are barely noticeable. People are wondering and expressing their affliction by staring at the devastation. There are images of people injured with superimposed text asking for a ceasefire, and other that show people just standing or walking in the middle of the ruins, rescue personnel extracting bodies from the debris. None of those images contains the visual elements referring to the causes of such destruction and suffering.

The real threat to people's lives remains invisible. This is not only surprisingly absent in the pictures where destruction is shown – as the one above – but even more in those focusing on the victim, the threatened subject that I have discussed in the previous chapter. The perpetrator, as "the victim's semiotic antibook" (Kurasawa 2015, 16), is never represented, either physically or symbolically. In visual social semiotics, the depiction of actions that only shows the goal, are called *events*: "something that is happening to someone, but we cannot see who or what makes it happen" (Kress and Van Leeuwen 1996, 64). Images in which the prime mover of the action is omitted or made anonymous are very similar, as Kress and Van Leeuwen (1996) have pointed out, to the linguistic construction of the *passive agent deletion*. This linguistic strategy is indeed considered in Critical Discourse Analysis as "ideological transformations": a grammatical maneuver considered as "containers of potential traces of ideological mystification" (Van Leeuwen 2009, 3). Critical Discourse Analysis has very effectively shown the difference between two different ways of accounting for the same event, based on alternative grammatical constructions. Van Leeuwen uses the example illustrated by Tony Trew in an article that become seminal (Trew cited in Van Leeuwen 2009). In 1975, the police in the white minority regime of Rhodesia (today's Zimbabwe) shot unarmed people, killing 13 of them. While the *Rhodesia Herald* led with the headline "A political clash has led to death and injury" the *Tanzanian Daily News* wrote how "Rhodesia's white supremacist police...opened fire and killed thirteen unarmed Africans." The study highlighted how two different political views had been encoded through choices regarding vocabulary and grammatical structure.

The visual absence of bombing, air raids, soldiers or weapons in the representation of devastation, suffering and victimhood works as a visual equivalent of the passive agent deletion. The power of the pictures above, and the many others of this kind, in denouncing the level of destruction, and in showing the kind of visual landscape Syrians on the move are living in when they embark on their migration journey, is therefore seriously undermined by the absence of the actor, the perpetrator. As a visual equivalent of the linguistic strategies that omit certain grammatical elements, these absences cannot only be considered as different stylistic choices, but rather as framings that encode specific political implications.

Although humanitarian NGOs' primary concern is not necessarily to inform about a situation – as is the case with the media – but, rather, to mobilize support, it is more difficult to attribute to these pictures a conscious omission. The visual absence of a threat has, rather, to be attributed to the NGOs' intentions to present themselves and their action as neutral and apolitical (as in the case of CARE, Oxfam and Save the Children as discussed in Chap. 4). It is more difficult to explain than in the case of MSF where the importance of calling out perpetrators and taking clear political stances cannot be reconciled with this specific invisibility. However, despite the reasons behind it, the visual absence of the threat contributes to a securitized depiction in so far as it overlooks, when not totally concealing, the causes of Syrian displacement. Also, when looked at through the lenses of the humanitarian perspective – to raise awareness of the plight of refugees and mobilize support – the lack of representation of what is causing people's suffering and flight fails to defy mainstream accounts in the name of alternative narratives.

7.2.3.2 Top Protection Issues

During the multi-sited fieldwork, I carried out for this study, I repeatedly asked NGO informants what were their current top advocacy messages. Although some topics would appear as priorities across countries, in each country – Lebanon, Jordan and Greece – the agencies had different sets of key concerns.

In Lebanon, NGO staff very often mentioned the need to target, not just exiled Syrians, but also vulnerable hosting communities (and Palestinian refugees) in the emergency response. This was due to assessments of similar high levels of need among vulnerable Lebanese and Palestinian host communities in addition to increasingly ever more explicit requests from the Lebanese government to include local communities given the rise in social tensions in the country.[11] When I asked an Oxfam adviser if the agency would visually present assistance provided to host

---

[11] Interviews with CARE Communication and Advocacy Manager, Beirut, 3rd March 2017 and Oxfam GB Lebanon Policy Advisor, Beirut, 8th March 2017.

communities the answer was emphatic: "No, because in any case, we are here to work on the needs and the rights of the most vulnerable". She elaborated that the point was not to think about the nationality of the beneficiaries but about levels of needs and vulnerability.[12]

Another important topic that often came up during interviews was the ongoing political debate around safe zones following a proposal by the Lebanese Minister for Foreign Affairs to set up refugee camps in areas of Syria that were considered safe, the so-called 'safe zones'.[13] On this point, one interviewee told me that "we are trying to develop a counter argument on the mainstream narrative on security zones in Syria. Everybody talks about security zones. Oxfam is saying that there are not safe zones and tries to develop a counter message". When I asked what the organization's communication strategy was to accomplish this objective, she answered: "we have done research on the refugees' perception of displacement and how they see their future. The research highlighted that they want to return but they want a safe return. We want to make their voices heard. We do that through photos, case studies, policy papers. We use a lot of quotes, to bring in the human element."[14]

Another top concern of humanitarian actors in Lebanon was the question of residency permits for Syrian refugees.[15] The question of documentation was, as they explained, crucial, as access to basic services provision such as health care, education, and food distribution hinged on having a valid residency permit. For refugees, obtaining one was extremely difficult and expensive. As the Lebanon Humanitarian INGOs Forum Advocacy and Communication Coordinator explained: "This is a very sensitive issue in Lebanon because the domestic political narrative is all about the fear that refugees will stay in Lebanon and residence is connected with permanency and not temporariness."[16]

Comparing these different priorities with the elements present in the NGOs' visual material brings out the striking inconsistency between the top humanitarian themes and the ones captured in the photos. Here is no visual reference to the neighborhood approach or to the positive stories countering the social tension narrative. For there is nothing about the safe zones debate. This is especially relevant if one considers that the numerous images showing the level of destruction occurring in Syria discussed in the previous paragraph, they all focus on areas where the territory was highly contested and was not considered potential 'safe zones'. Similarly, I did not find visual mention of the controversial topic of residence permits and the daily implications for Syrian refugees of lacking valid documentation.

---

[12] Interview with Oxfam GB Lebanon Policy Advisor, Beirut, 8th March 2017.

[13] For an analysis of the debate and the political implications of safe zones in Syria see Atallah and Mahdi 2017 and Dionigi 2016.

[14] Interview with Oxfam GB Lebanon Policy Advisor, Beirut, 8th March 2017.

[15] Interviews with Save the Children Media Advocacy and Communications Manager, Beirut, 8th March 2017, and Lebanon Humanitarian INGOs Forum Advocacy & Communications Coordinator, Skype, 15th March 2017.

[16] Interview with Lebanon Humanitarian INGOs Forum Advocacy and Communications Coordinator, Skype, 15th March 2017.

In Jordan, NGO staffers listed an even wider set of humanitarian concerns. Interviewees often mentioned how closure of the Syrian border was affecting the right of people to seek asylum, the medical evacuation of war wounded and worsening the dramatic humanitarian situation of people stuck at the berm (barrier of earth and sand) between Syrian and Jordan. They highlighted how serious human rights violations were happening due to non-respect of the principle of *non-refoulement*[17] and increasing use of deportation.[18] Regarding the situation of refugees already residing in the country, interviews with humanitarian practitioners confirmed Achilli's analysis of the Jordanian Government policy of encampment implemented since 2014: "A bleak scenario is playing out against the backdrop of the Jordanian government's new policies. The encampment policy has affected most Syrian refugees in Jordan at three interrelated levels: it has shrunk the humanitarian space and raised considerable protection concerns; it has increased the number of ITS [Informal Tented Settlement – the default option for refugees to reside in camps or unable to afford housing solutions within host communities] evictions, refugee deportations to camps, and refoulement to Syria; and it has forced refugees into negative coping mechanisms" (Achilli 2015, 6).

Humanitarian informants reported increasing difficulty in accessing the documentation needed to legally reside in urban areas of Jordan and access basic services. An NGO mission head reported that: "The question of valid documentation has become a priority concern. Since 2014, in order to access the Ministry of Interior (MOI) cards that allow the provision of services outside the camps, where 80 per cent of the Syrian refugee population lives, people have to go through a long and expensive bureaucratic process that leaves many unable to obtain the cards, either because they do not meet all the criteria (valid documentation from the camps, written and certified rental agreement, etc.), or because they cannot afford the medical certificates needed for all members of the family. The result is that if you do not have the MOI card you cannot enroll your children in school, access health facilities, receive food or non-food item assistance from UNHCR and other UN agencies."[19]

The photographic account of Syrian refugees in Jordan does not correspond with the set of humanitarian concerns spelled out by NGOs practitioners. Images of people stuck at the berm have rarely appeared in the media, with no mention in visual materials of relief agencies. NGOs have sometimes published statements pointing to the protection concerns arising from closure of the Syrian border, but text has not

---

[17] As the *United Nations Declaration on Territorial Asylum* unanimously adopted by the General Assembly in 1967 affirms: "No person referred to in Article 1, paragraph 1, shall be subjected to measures such as rejection at the frontier or, if he has already entered the territory in which he seeks asylum, expulsion or compulsory return to any State where he may be subjected to persecution" (UN General Assembly 1967).

[18] Interviews with MSF Humanitarian Affairs Officer, Amman, 17th May 2017, Intersos Regional Protection Advisor, Amman, 14th June 2017, Syria INGOs Regional Forum (SIRF) Coordinator, 13th June 2017, Jordan INGOs Forum (JIF) Advocacy Officer, 17th May 2017.

[19] Interview with Intersos Head of Mission in Jordan, Amman, 25th June 2017.

been accompanied by images. There have been no images of evictions, deportations, forced return to camps, difficulties in obtaining required documentation or limits on access to basic services. Clearly, NGOs have limited space for maneuver to publicly disseminate images of actions that the host government is carrying out in exercising its sovereignty. Nevertheless, the omission is striking, especially considering how some of the NGOs consider themselves champions in human security protection, even if this involves speaking out against states.

Two factors help explain the complete invisibility of key protection issues in Lebanon and Jordan. On the one hand the NGOs (at least in the case of CARE, Save the Children and Oxfam) are keen to present themselves as neutral and apolitical actors, focusing only on what they insist are 'technical' humanitarian aspects, such as human rights or provision of services, and eschewing reference to themes with potential political implications. Such a strategy is short-sighted, making it less likely to realise their 'technical' goals, and at the same time most likely one of the main reasons of this visual absence. On the other hand, the agencies focus on intense advocacy with the Government on protection concerns. As an MSF staffer reported: "With the political context you have to work taking into consideration that for the benefit of the beneficiaries you can never stand up too strongly because you risk getting the opposite effect. In particular for public communication, we always analyze the benefits and risks, and analyze if we are compromising our or other operations. Often the political context blocks advocacy, public communication and therefore we only follow private advocacy. There are cases for example where we cannot continue. We would like to engage through public communication initiatives, but it is better not to because at the same time we were following private negotiations with the government."[20]

Clearly in Jordan and Lebanon, as in many refugee hosting contexts across the globe, NGOs have to frequently assess communication and public advocacy against a cost-benefit analysis of threats to their wider in-country operations. This caution is probably another important factor in trying to understand the reasons behind this aspect of invisibility. All this notwithstanding, the visual absence of protection concern remains a significant gap in humanitarian communication and particularly with regards to NGOs' objectives of raising awareness and mobilize support on the condition of displacement that, in this way, remain largely invisible.

In Greece, Syrian refugees were facing a different set of protection concerns, as an MSF manager explained: "First, the need for safe passages and legal passages. I am not talking about the opening of borders but official safe routes. Second, the EU needs to assume its responsibilities. We publicly took a position against the EU-Turkey deal with our publication *One year from the EU-Turkey deal: challenging the EU's alternative facts*. Third, the fact that camps are not a solution. Then, on a secondary level, we also have to address refugees living conditions and wellbeing."[21]

---

[20] Interview with MSF Humanitarian Affairs Officer, Amman, 17th May 2017.

[21] Interview with MSF Director of Operational Support Unit Athens, Athens, 25th April 2017.

The impact of the EU-Turkey deal on Syrians on the move was, as noted earlier, frequently mentioned by NGO interviewees. As a Save the Children communications specialist in Greece told me: "Currently one of the key messages is around the EU-Turkey deal, on the impact that the deal has on the kids especially on the islands. Their psychosocial status is increasingly worse with more and more attempting suicide and harming themselves. We are calling for the EU to improve conditions on the islands, take families and children out of the detention centers, give them a fair right to asylum and that every European country does it fair share for resettlement. (…) We are also putting pressure on the government to remove children and families from detention centers and send them to Athens. The problem is that this would mean a breach of the EU-Turkey deal according to which the refugees should either stay on the islands or be sent back to Turkey."[22]

Beside critique of the EU-Turkey deal, NGO staffers also mentioned other concerns for mobile Syrians. Everybody agreed that the top priority was the situation in particular camps such as Moria on Lesbos where most residents were single men, encaged behind tall fences with lack of activities and additionally the presence of minors in a volatile security environment.[23] The case of Greece is particularly interesting because the journey of Syrians on the move through the Balkan route into central Europe is probably not only the best visually documented, but also the one that best reflects border closures, the harsh conditions faced by refugees and protection risks. NGO images of this movement include a much wider range of visual themes than found in the way refugees in Syria's neighbors are represented. At the visual level there appears to be more coherence with the key protection concerns highlighted by humanitarian practitioners.

The reason behind this exception could be that in the "economy of attention" (Citton 2014) –in which NGOs like all communication actors compete – the *space* available has been wider when the so-called refugee emergency impacted Europe compared with humanitarian conditions in Turkey, Jordan and Lebanon that remained more distant for the (mostly European) Western audience. This highlights, again, the differential attention which humanitarian agencies give to contexts of need in different parts of the world. It also shows that the representation of displacement can have a much wider set of visual themes than the threat/threatening dichotomy which we are used to. The visual elements of borders, fences, restricted mobility, law enforcement, corpses and coffins, contribute to a depiction of the situation that is more complex and therefore closer to conveying actual displacement conditions. Instead of undermining the main argument of this chapter – the importance of presence and absence – this confirms that the dialectic between visibilities and invisibilities is crucial in better understanding the potential for NGOs' visual

---

[22] Interview with Save the Children, Information, Communication and Media Manager, Athens, 24th April 2017.

[23] Interviews with Solidarity Now Executive Director, Athens, , 27th April 2017; MSF Coordinator, Lesbos, 2nd May 2017; Karatepe Camp Manager Assistant, Lesbos, 2nd May 2017; Save the Children Child Protection Manager, Lesbos, 3rd May 2017; Praxis protection team, Lesbos, 6th May 2017.

material in terms both of oppression or emancipation. The power of (in)visibilities is exactly that of further concealing, or better highlighting, the humanitarian, political and security implications that a specific situation has on the lives of the people involved.

7.2.4 Refugees Voices: Silence and Invisibility

Make people's voices heard! This is an imperative for most of the agencies here studied. It constitutes a key point of action. Save the children's advocacy strategy is directed toward two objectives. The agency "advocates and campaigns for change to realize children's rights and to ensure that their voices are heard" (Save the Children 2017). In a strategy document the agency affirms it will always "be the voice: advocate and campaign for better practices and policies to fulfill children's rights and to ensure their voices are heard (particularly most marginalized or those living in poverty)" (Save the Children 2016, 3). Oxfam's Strategic Plan 2013–2019 affirm that its vision "sets local communities and the voices of women, men and young people at the centre of change" (Oxfam 2013, 5) and that the organization "will put a particular focus on gender justice and empowering poor people to make their voices heard" (Oxfam 2013, 8). "Oxfam believes that people living in poverty who claim their rights and make their voices heard constitute an enormous source of hope for real change and greater power in people's lives" (Oxfam 2013, 9). Along similar lines, CARE declares that the organization puts "empowering women and girls at the centre of what we do, providing opportunities for their advancement and ensuring their voices are heard" (Care International 2017, 8). By contrast, MSF does not mention such a programmatic objective in any of its official strategic documents. Only one of the agency's publications – entitled *Hear my voice* (MSF OCA 2012) – makes reference to this to this rhetoric. MSF's emphasis on speaking out publicly when it witnesses extreme violence or when crises are neglected, marks it out, as noted earlier, from the other agencies studied. Save the Children, Oxfam and CARE explicitly state their role in making people's voices heard, while MSF retains the option to raise its own organizational voice. The emphasis given by humanitarian organizations to this programmatic objective and the fact that it constitutes a common strategy, indicates its centrality in humanitarian discourse and aspiration to stand in solidarity with the most vulnerable and marginalized. But it is also part of NGOs' attempts to challenge securitized accounts of refugees by conveying individual stories and highlighting the human dimension of displacement through people's own voices. Against this background, I believe it is crucial to explore the value and meaning of invisibility and silence.

A note here is in order here to clear way the confusion that may be caused by this apparently strange synesthesia.[24] From an analytical perspective, I am equating

---

[24] Synesthesia is a figure of speech through which the evocation of one sense is linked with a predicate evoking of another. For example: "the silence that dwells in the forest is not so black" (Wilde 2015), in which the visual attribute of color is attributed to the auditive sense of silence.

silence with invisibility and having one's voice heard with visibility. It is no coincidence that in a blog post on Oxfam websites' policy and practices section, the two dimensions are mixed up. For the organization Gender and Governance Adviser in fact, raising women's voices means 'making women's invisibility, visible' (OXFAM 2017). More generally, silence and invisibility have been treated in academic literature in the different disciplines as two sides of the same coin (e.g. Dimitrov 2015; Sparkes 1994; Simms 1986; Kim 1998; Coombs et al. 2014). It has probably have to do with the fact that the two dimensions are associated with the politics of hiding (Jones et al. 2016). Indeed, both notions have to do with absence and presence and visual practices can be used ambivalently to empowering and give voice, or on the contrary, to conceal or silence specific groups of people. This section is focused on exploring how absence, invisibility and silence constitute the humanitarian subject and how this representation resonates with the securitization of Syrian refugees.

Reflecting on silence and securitization, Lene Hansens' article on *The Little Mermaid's Silent Security Dilemma and the Absence of Gender in the Copenhagen School* (2000) addresses gender in securitization theory. Leaving aside her discussion on gender – or, rather, the lack of consideration of it in the Copenhagen School –her considerations on silence are particularly relevant for this analysis. Hansen argues that the framing of a particular issue in terms of security cannot occur in situations where the referent object is silent or silenced. I will return to this point in my discussion of 'silence' in the visual material. Hansen calls for an epistemological inclusion of non-verbal forms of communication such as the visual. Following this logic, silence and silencing, exactly like absence and invisibility appear strictly interconnected with securitization processes. This is all the more evident if we move from a conception of securitization as a speech act, in the strict sense of the term, and embrace a wider notion of securitization utterances that includes the visual.

What is strikingly absent from NGOs' visual material is exactly the different voices, the opinions and positions of the people directly impacted by displacement and humanitarian assistance. As I have stressed above, relief organizations' visual communication has undoubtedly privileged a representation of people on the move highlighting their individuality and personal traits, a depiction that has moved away from photos of indistinguishable people that Malkki has defined as a "miserable sea of humanity" (Malkki 1996, 377). In almost every single case, the pictures are accompanied by captions, posts or quotations that give some insight into the personal trajectories of the people photographed. Although very often names are changed for reasons of confidentiality and security, almost all images are complemented by some textual information on who is the represented subject, how old s/he is, where s/he comes from and what is his/her problem.

There is an Oxfam portrait of a woman where every detail of her lineaments is in sharp focus (Oxfam 2016a). Behind her piercing stare, we can see her dark brown eyes, every wrinkle, underling the expression of gravity and pride of this older woman. We are so close to her that we can see the print of her scarf and the texture of the scarf that frames her lower jaw. The background is out of focus, in the typical style of portraits, where everything is about the person photographed. The caption indicates that this picture was been taken by a professional photographer, Sam Tarling, commissioned by Oxfam. The style, the choice regarding the depth of focus, the eye-to-eye angle of the shot, the perfect lighting, the direct gaze, everything draws attention to this woman, her individuality, the distinctiveness of her face and her gaze. Accompanying text gives us some information: "Amna from Deraa in Syria inside her caravan in Zaatari camp. When Amna and her family arrived in Zaatari they all had to sleep in just two tents with blankets but no mats to protect them from the stony ground below. Amna said there was no water and sanitation in her area of the camp until Oxfam installed them" (Oxfam 2016c).

The scope of this picture and the text is to offer the viewer some insight into the lives of Syrian refugees in Jordan. The article containing the image is entitled "From Syria to Zaatari: glimpses of refugees in Jordan." 'Glimpses' is a revealing choice of word, referring to the possibility of *seeing* how it is like, *looking* at the faces of the people affected by war and displacement. It is about visibility. However, despite the intensity of the portrait, we still do not know anything about this woman. How was her life in Syria? How is she doing in Jordan? All that we get to know is that she was living in an uncomfortable situation and that thanks to Oxfam conditions in the camp have improved. Her gaze, her powerful eyes, her wrinkled visage all hint at more. We expect her to start telling us what brought her from home in the southern Syrian city of Daara to Zaatari, a massive refugee camp outside the Jordanian city of Mafraq. But there is nothing more. She appears again elsewhere on Oxfam's website, this time looking away from the camera, her portrait used as the face of "Crisis in Syria" (Oxfam 2016b). This time, there is even less about her. She simply serves the purpose of encouraging us to click on the "donate" arrow to the left of her image.

Similar dynamics are at play elsewhere. Interestingly, a 2016 CARE photo gallery (entitled "Dear World. Photos Capture Syrian Refugees Voices" there are a series of single images with no accompanying background information. Delving deeper, I found that two American photographers, Robert Fogarty and Ben Reece, had first given their subjects marker pens and then encouraged them to write messages to world leaders on their arms and hands. Then Reece and Fogarty, who travelled with CARE to Jordan in 2013, photographed them' (CARE 2013). In the gallery, each image, exactly as in Fogarty's Dear World project,[25] is either a portrait of a single individual or a small group of children. All the represented participants have words written on their arms, messages to the world. For example, there is the picture of a girl, her face and hand badly burned, looking directly into the camera (and to the viewer). The caption gives us some information that explain the wounds on her face, results of the bombing of her house and lack of immediate health care (CARE 2016). As in the cases mentioned above, this portrait has a strong potential, but it is in a way limited by its use. Each person becomes a message. It is more about the slogan than about people's voices. Once again, I am not questioning the image

---

[25] A commercial photographic project that creates events around portrayals of people with a message written on their arms. For more information see http://www.dearworld.com/

per se, which in this case, as in many others of its kind, represents people with dignity and pride, thus far from the stereotypical victimization of the humanitarian subject. The problem of these representations is that they claim to represent people's voices. But the personal trajectories, the different lived experiences, the individual opinions and narratives of these people remain largely invisible. While it is quite evident that NGOs producing these images are attempting to empower the represented participants and to draw attention to 'human face' of war and displacement, there is so much that is left out of the picture that eventually the photographic act remakes them invisible.

They become visible only as humanitarian subjects while their real individuality, their lives, stories, trajectories, perspectives, and claims remain completely invisible. Most of the pictures focusing on the individuals affected by the Syrian war and displacement mention names, ages, maybe one sentence of information on where the person is living or which particular humanitarian programme is helping him/her. What is ultimately lacking in all these pictures is refugees' voices, genuine ones that cannot be simply reduced to a sentence of accompanying textual information. The messages, indeed, can be very powerful and useful in a communication-commercial perspective, just as much as words on mugs, T-shirts or other merchandise, can be. Clearly, this is important especially considering that NGOs have to fundraise to enable field interventions and that media campaigns are used to draw public attention to humanitarian situations.

What, however, is important is not to equate such photo projects and media campaigns[26] with providing refugees with an opportunity to have their voices heard. Buried behind hundreds of images portraying Ahmad, an 8-year-old from Aleppo, Du'a, a 5-year-old from Homs, or Huda, 38 year-old mother of two from Daraa, people remain invisible and the opportunities for them to speak and for us to hear them seriously limited.

As Bal has brilliantly put it, the "thing on display comes to stand for something else, the statement about it. It comes to mean. The thing recedes into invisibility as its sign status takes precedence to make the statement" (Bal 1996, 4). These kinds of images, representing single individuals or small group of people, with very sparse information seeking to 'put a name to a face' constitute a dominant visual theme of present-day humanitarian communication. Whereas they have helped the shift from preoccupation with a 'sea of humanity' to engaging with individuals affected by the crisis, they have not managed to give people a voice or make them individually noticeable, beyond their likeness in communication material. The possibilities of expression are evidently limited by the communication format that creates the space for people's voices: there is room for one sentence on the arm, one object to

---

[26] See, for example, CARE campaigns #not that different or 2000 days available at: https://www. care-international.org/news/photo-video-gallery/photo-gallery or SCF's Facebook campaign *Do not let their future disappear* (https://www.facebook.com/savethechildrenuk/ photos/a.419080615190.375445.117476785190/10156504811555191/?type=3), and *Every Last Child* (https://www.savethechildren.org/us/more-ways-to-help/campaign-with-us/ every-last-child)

symbolize the experience of loss, one phrase to describe what they miss the most after years of displacement. These visual projects have great potential from a communication and fundraising perspective. Thanks to the combination of various semiotics resources at play (including direct gaze, eye-level angle, close-up shot and subtle lighting) they do spark an emotional connection between the viewer and the represented participants. By evoking engagement and a close social relation, the photographic choices about contact, distance, perspective, and angle reinforce the request for our involvement with the subjects of the image. The potentiality of these pictures to mobilize attention and humanitarian response are not in question. What is vital is to realize, however, is that they are not a means through which to convey the variety of people's voices, opinions and claims. It is misleading and unjust to assume Syrians' voices can be represented by such photographic projects.

It is important to observe that not all images are like those presented above. In some cases, there were more than one picture representing each individual personal story and the text provided more information on a specific person or family, an example is the Save the Children article, *Layla and Sana sleep without fear of snakes* (Save the Children 2016). The article and pictures show how Layla's and Sana's family have faced prolonged difficulties, most especially with regard to shelter. The piece describes the father's challenges in finding a job (but again without entering into the thorny issue of how host states erect obstacles to any kind of income generating activity), the lack of proper accommodation that caused their mother to miscarry and the children's fears. The article concludes with information how the agency has intervened to provide better shelter.

In this case, as in many others the focus is the intervention of the organization and how this has improved the situation of the represented participants. Both at the visual and textual level we cannot understand anything more about their situation, beyond their humanitarian need. The focus is not really about who we see in the pictures, certainly not about their voices: rather, it is about what NGOs are doing. People remain consistently invisible and what we receive is an abstract humanitarian 'subject'. This hyper-visibility of the humanitarian subject happens at the expenses of the possibility of visibility for individuals with their varied expressions, narratives and experiences.

Looking at this (in)visibility through the lenses of securitization theory, it seems that contemporary visual representations of Syrians, while aiming to make people's voices heard, continues only to underline their humanitarian needs and silence any other individual and collective claim. In this sense, NGOs' communication reproduces a narrative in which the person is 'only' the referent object of a threat in need of external intervention. Any other claim is silenced.

In a ground-breaking article on the universalizing depiction of people on the move as "refugees", Malkki (1996) has argued that: "One of the most far-reaching, important consequences of these established representational practices is the systematic, even if unintended, silencing of persons who find themselves in the classificatory space of 'refugee'. That is, refugees suffer from a peculiar kind of speechlessness in the face of national and international organizations whose object of care and control they are. Their accounts are disqualified almost *a priori*, while the languages of refugee relief, policy science, and 'development' claim the production of narratives about refugees" (Malkki 1996, 386).

Her analysis is primarily based on representation of refugees escaping genocide in Burundi in 1972 in Rwanda in 1994, typically depicted as speechless, universal victims, masses of bodies. As I have noted before, present-day humanitarian communication has put a great deal of effort into moving away from such visual representation. However, this 'new' representation notwithstanding, they remain largely speechless. What we hear is seriously limited by the NGO's intermediary role.

In trying to find a solution for the 'little mermaid security dilemma', to give voice to those affected by crisis, NGOs mediate this voice. The problem is that while trying to address the silence of vulnerable people NGOs are re-silencing them by the manner in which they seek to speak on their behalf. As Fassin has insightfully pointed out: "Of course, the humanitarian agents who collect accounts or carry out inquiries to reveal the violence or injustice suffered by oppressed or displaced or bombed populations base their testimony partly on what the victims of this violence or these injustices say about them. Their third-person testimony is grounded on first-person testimonies. However, the requirements of defending causes and the logic of their intervention lead them to what might be termed a humanitarian reduction of the victim. On the one hand, all that is retained of people's words is what contributes to a telling image in the public space. (...) On the other hand, the individuals in question tend to conform to this portrait, knowing that it will have an impact on public opinion, and thus offer to the humanitarian agents the part of their experience that feeds the construction of them as human beings crushed by fate. (...) It is rather that if one believes that what distinguishes humans from other living beings is language and meaning and that what makes human life unique is therefore that it can be recounted, as Hannah Arendt asserts, then humanitarian testimony establishes two forms of humanity and two sorts of life in the public space: there are those who can tell stories and those whose stories can be told only by others" (Fassin 2007, 517).

Refugees' voices are invisible or unheard because they get lost in the process of NGOs communication production. Rajaram has noted how "bureaucratization of knowledge" (Rajaram 2002, 248) and the mediation of refugee's experience in making the refugees' voices heard, fails to problematize and acknowledge the presence of the author and, most importantly, her/his position. By investigating an Oxfam project entitled *Listening to the Displaced*, he concludes that humanitarian representation eventually "consigns refugees to their bodies, to a mute and faceless physical mass" (Rajaram 2002, 247) with the ultimate consequence of silencing them and commodifying their experiences.

Several dynamics that Rajaram highlights are very similar to the ones at play in the visual representation of Syrian displacement. As he notes, through the process of leveling and abstracting refugee voices to make everyone's voices heard – including those in vulnerable categories of children, women and older persons – NGOs downgrade the compound nature of people's identities. They purport to aim to allow people to speak for themselves yet NGOs do not question the general structure and politics of the humanitarian enterprise, with the result that the only narration around displaced people is the one that fits with humanitarian needs and the relevance of relief intervention. In fact, as Solomon-Godeau has argued, "dominant social relations are inevitably both reproduced and reinforced in the act of imaging those who do not have access to the means of representation themselves" (Solomon-Godeau and Nochlin 1991, 180). The text and images here examined may be poignant, powerful elicitors of sympathy. However, they are deprived of politics and context, confining refugees to their essentialized bodies "invariably detached from the local historical context of the reality that they supposedly represent" (Escobar 1999, 108 quoted in Rajaram 2002, 256).

Any form of self-representation is strikingly absent in the contemporary high-tech media and social media landscape within which NGOs disseminate visual material on humanitarian crises. The complete denial to refugees of opportunity to represent themselves, has mostly remained overlooked by academia with a few notable exceptions (Chouliaraki 2017; Literat 2017; De Leeuw and Rydin 2007). Of relevance is anecdotal experience recounted by a professional photographer who has worked in Jordan throughout the refugee crisis. In 2016, she was commissioned by Save the Children teach photography to young people in Zaatari camp: "I gave an assignment to my students: to take pictures to tell us about their life as refugees in Jordan, and I asked them to write captions to accompany the photos. They did a tremendous job but what impressed me the most was the fact that the captions completely inverted the sense we are used to giving to pictures produced in the context of a refugee camps. For instance, I remember that a smiling girl with a red stuffed animal (that make me immediately think of the photographic trope of portraits of children with a favorite toy). The caption said: *my sister is smiling because we always fight over this stuffed animal and she finally managed to win it this time* (instead of the poor Syrian refugee who could not save anything else when she fled her house). Similarly, I have seen pictures taken by people cooking a good meal on the occasion of a holiday and the caption: *today is a great day, we get together, we eat meat, we will enjoy.*"[27] When I asked her where I could see these pictures and the results of the photographic project, she told me that, sadly and despite all her efforts, the agency did not want to organize an exhibition and, as far as she knew, never used any of the photos in communication materials.

This story is paradigmatic in showing how refugees' self-representation could add a variety of new meanings and visual narratives to the ones typically (and stereotypically) used in present-day humanitarianism. Sigona has pointed out how the diversity and plurality of refugees' voices "does not necessarily translate into humanitarian, academic, and media discourse, as these tend to privilege a one-dimensional representation of the refugee which relies heavily on feminized and infantilized images of 'pure' victimhood and vulnerability" (Sigona 2013, 370).

---

[27] Interview with a professional photographer who had extensive experience working with NGOs and refugees in Jordan, Amman 26th May 2017.

7.3   Conclusion

This chapter has explored the value of absence. It has done so with an eye on the securitizing framework typical of NGOs' representation of displacement both in terms of human security and in their attempt to challenge media and populist discourses that depict refugees as threat. It has shown that topics or events that can be very relevant to the lives of people living in displacement or on the move tend to be left aside by NGOs in their representation of the humanitarian emergency. Symbolic and non-symbolic references to politics such as signs of victory or display of national or factional flags are completely absent from relief organization's visual repertoire. This absence seems to be linked with NGOs' explicit intention to not implicate themselves in local conflict politics, to keep their humanitarian work as distant as possible from any perception of political involvement of either the organization or of the beneficiaries.

At the same time, the absence of politics is also a way to protect assisted refugees whose relatives inside Syria could be at risk if they were thought to be affiliated with protagonists. Protest is absent from NGOs' visual communication. Contestation and protests by people on the move may sometimes be represented in media accounts but not in those generated by relief organizations. They most likely exclude protests in order to not unnecessarily fuel securitized discourses that depict refugees in terms of threat. However, the absence of the visual themes of politics and protest has also the effect of silencing the voice and the political agency of refugees, eventually and perversely doing the opposite, silencing their political agency and depoliticizing refugees.

The third section of the chapter investigated the visual presence/absence of the threat to people on the move. When refugees are represented as threatened, the way the threat to their lives is depicted becomes relevant, especially in a securitizing framework in which for the securitizing move to start the referent object needs to be presented at serious risk of a specific set of threats. Regarding the threats in Syria, the analysis has shown that most of the pictures represented war and its effects through images of destruction: bombed buildings, ruins, rescue personnel frantically plucking people from the rubble. Such images powerfully chronicle the level of devastation yet consistently leave out any representation of the perpetrators – whether as the Syrian military, militias, weapons, tanks or those who direct aerial bombardments. Their visual absence, the aesthetic equivalent of the grammatical stratagem of 'passive agent deletion', has the ultimate effect of concealing the perpetrator, with all the implications of such an ideological move.

Moving from threats inside Syria to threats in host countries or those through which refugees seek to pass, the investigation of the visual presence of the top protection issues mentioned by humanitarian practitioners in each country visited during fieldwork has revealed a more complex situation. In Lebanon and Jordan analysis has demonstrated a strong disconnection between the topics that NGOs' staff identified as priority concerns and the visual themes present in the images. Basically, none of the topics raised by humanitarian practitioners was included in the NGOs' visual material. Exploring key humanitarian concerns in Greece and, more generally those arising from the arrival of Syrians in Europe, the analysis has revealed a much tighter connection between the visual and the protection discourse with images presenting the set of visual topics considered to be top humanitarian priorities.

This chapter has also explored the presence/absence of refugee voices in the humanitarian communications, highlighting the superficiality of introducing specific individual stories. Often, the only insight we (as audience) get to know about the lives of those portrayed is their name and age. Even when more details are provided on individuals, the representation is limited to their humanitarian need with any other aspect constituting the person, beyond his/her humanitarian subjectivity, is often overlooked. It is as if the hyper-visibility of the human face of the humanitarian crisis is only possible at the expense of the visibility of individual, remain mostly invisible and undistinguishable from others. It is in the mediation role that NGOs undertake in their attempt to 'make refugees voices heard', that their voices get eventually lost. The analysis has also shown how any form of self-representation by people on the move is left out from the organization's communication material. The effect is to leave NGOs as the only actors enabled to speak or convey the refugees' voices, inevitably reducing them to their humanitarian vulnerability and victimhood through the mediation process.

The investigation has also shown that both visual presence and absence can reinforce an emancipatory or oppressive message. This is particularly evident in the case of the (in)visibility of politics and protest where the absence of these visual themes simultaneously works toward the depoliticization but also the (securitized) protection of the people on the move. This chapter, and the last section in particular, has highlighted how the visibility of a wider set of visual themes on a specific issue (e.g., not only the basic humanitarian need faced by Syrians on the move seeking refuge in Europe, but also fences, borders and law enforcement agencies) could potentially offer a more complex and nuanced depiction of displacement in its wider political context. In short, the more that is shown about the historical and political situatedness of a crisis, the less people affected by it are described only in terms of threatened victims in need of humanitarian protection. The value of visibility, in the end, is therefore in being able to confer more agency to the people represented.

References

Achilli, L. (2015). *Syrian refugees in Jordan: A reality check.* Policy Brief, MPC, EUI. http://cadmus.eui.eu/bitstream/handle/1814/34904/MPC_2015-02_PB.pdf

Al Jazeera. (2015, November 23). *Refugees sew lips in Greece-Macedonia border protest.* https://www.aljazeera.com/news/2015/11/refugees-hunger-strike-greece-macedonia-border-151123152724415.html

Atallah, S., & Mahdi, D. (2017). Law and politics of "safe zones" and forced return to Syria: Refugee politics in Lebanon. *The Lebanese Center for Policy Studies, 39,* 36.

Bal, M. (1996). *Double exposures: The subject of cultural analysis.* New York: Psychology Press.

Bleiker, R., Campbell, D., Hutchison, E., & Nicholson, X. (2013). The visual dehumanisation of refugees. *Australian Journal of Political Science, 48*(4), 398–416.

Butler, J. (2004). *Precarious life: The powers of violence and mourning.* London: Verso.

Campbell, D. (2007). Geopolitics and visuality: Sighting the Darfur conflict. *Political Geography, 26*(4), 357–382.

Campbell, D., & Shapiro, M. J. (2007). 'Guest editors' introduction – Special issue on securitization, militarization and visual culture in the worlds of post-9/11. *Security Dialogue, 38*(2), 131–137.

CARE. (2013). *"Dear world": Photos capture Syrian refugee voices.* Text. CARE. http://www.care.org/emergencies/syria-crisis/dear-world-photos-capture-syrian-refugee-voices

CARE. (2016). Http://Www.Care.Org/Gallery/1179. Text. CARE. http://www.care.org/gallery/1179

Care International. (2017). *Care International annual report FY16.* https://www.care-international.org/files/files/publications/Care_International_Annual_Report_2017_8921_ENG_12_Online(4).pdf

CARE Syria Response. (2016). *CARE Syria response – Posts.* https://www.facebook.com/CARESyriaResponse/photos/a.362149093938941.1073741828.361464940674023/78874 1501279696/?type=3&theater

Carter, K. (1993). *The struggling girl. 100 photographs. The most influential images of all time.* http://100photos.time.com/photos/kevin-carter-starving-child-vulture

Chouliaraki, L. (2017). Symbolic bordering: The self-representation of migrants and refugees in digital news. *Popular Communication, 15*(2), 78–94.

Citton, Y. (2014). *L'économie de l'attention: Nouvel Horizon Du Capitalisme?* La Découverte.

Clarke, K. (2017, March). Protest and informal leadership in Syrian refugee camps. In *Refugees and migration movements in the Middle East* (POMEPS studies). https://pomeps.org/wp-content/uploads/2017/03/POMEPS_Studies_25_Refugees_Web.pdf

Coombs, D., Park, H.-Y., & Fecho, B. (2014). A silence that wants to be heard: Suburban Korean American students in dialogue with invisibility. *Race Ethnicity and Education, 17*(2), 242–263.

De Leeuw, S., & Rydin, I. (2007). Migrant children's digital stories: Identity formation and self-representation through media production. *European Journal of Cultural Studies, 10*(4), 447–464.

Dimitrov, R. (2015). Silence and invisibility in public relations. *Public Relations Review, 41*(5), 636–651.

Dionigi, F. (2016). *The Syrian refugee crisis in Lebanon: State fragility and social resilience.* London: Middle East Centre, LSE.

Ericson, R. V., & Haggerty, K. D. (2006). *The new politics of surveillance and visibility.* Toronto: University of Toronto Press.

Falk, F. (2010). Invasion, infection, invisibility: An iconology of illegalized immigration. In *Images of illegalized immigration: Towards a critical iconology of politics* (pp. 83–100). Bielefeld: Transcript.

Fassin, D. (2007). Humanitarianism as a politics of life. *Public Culture, 19*(3), 499–520.

Foster, H. (1988). *Vision and visuality.* Seattle: Bay Press.

Foucault, M. (1975). *Discipline and punish: The birth of the prison.* New York: Random House.

Franklin, A. J. (1999). Invisibility syndrome and racial identity development in psychotherapy and counseling African American men. *The Counseling Psychologist, 27*(6), 761–793.

Hansen, L. (2000). The little mermaid's silent security dilemma and the absence of gender in the Copenhagen school. *Millennium, 29*(2), 285–306.

Huysmans, J. (1995). Migrants as a security problem: Dangers of "securitizing" societal issues. In R. Miles & D. Thranhardt (Eds.), *Migration and European integration: The dynamics of inclusion and exclusion.* London: Pinter Publishers.

Jones, R. D., Robinson, J., & Turner, J. (2016). *The politics of hiding, invisibility, and silence: Between absence and presence.* London: Routledge.

Kim, J. H. (1998). *Silence and invisibility: A feminist case study of domestic violence in the lives of five Korean-American women*. Urbana: University of Illinois at Urbana-Champaign.

Kleinman, A., & Kleinman, J. (1996). The appeal of experience; the dismay of images: Cultural appropriations of suffering in our times. *Daedalus, 125*(1), 1–23.

Kress, G. R., & Van Leeuwen, T. (1996). *Reading images: The grammar of visual design*. London: Psychology Press.

Kurasawa, F. (2015). How does humanitarian visuality work? A conceptual toolkit for a sociology of iconic suffering. *Sociologica, 9*(1). https://doi.org/10.2383/80396.

Leander, A. (2016). Digital/commercial (in) visibility: The politics of DAESH recruitment videos. *European Journal of Social Theory*. https://doi.org/10.1177/1368431016668365.

Literat, I. (2017). Refugee selfies and the (self-) representation of disenfranchised social groups. *Media Fields Journal, 12*, 1–9.

Malkki, L. H. (1996). Speechless emissaries: Refugees, humanitarianism, and dehistoricization. *Cultural Anthropology, 11*(3), 377–404.

Manzo, K. (2008). Imaging humanitarianism: NGO identity and the iconography of childhood. *Antipode, 40*(4), 632–657.

MSF OCA. (2012). *Hear my voice. Somalis on living in a humanitarian crisis*. http://www.msf.or.jp/library/pressreport/pdf/1302_130212_HearMyVoice_MSF_OCA.pdf

Oxfam. (2013). *The power of people against poverty. Oxfam Strategic plan, 2013–2019*. https://www.oxfam.org/sites/www.oxfam.org/files/oxfam-strategic-plan-2013-2019.pdf

Oxfam. (2016a). *Amna from Deera – From Syria to Zaatari Glimpses of refugees in Jordan*. https://web.archive.org/web/20181004153253/https:/www.oxfam.org/en/syria-zaatari-glimpses-refugees-jordan

Oxfam. (2016b). *Emergency response*. Oxfam International. https://www.oxfam.org/en/emergencies

Oxfam. (2016c). *From Syria to Zaatari Glimpses of refugees in Jordan*. Oxfam International. https://www.oxfam.org/en/syria-zaatari-glimpses-refugees-jordan

OXFAM. (2017). *Raising her voice: Making women's invisibility, visible*. Oxfam GB, Policy & Practice. http://policy-practice.oxfam.org.uk/blog/2014/06/raising-her-voice-making-womens-invisibility-visible

Rajaram, P. K. (2002). Humanitarianism and representations of the refugee. *Journal of Refugee Studies, 15*(3), 247–264. https://doi.org/10.1093/jrs/15.3.247.

Rancière, J. (2014). *The emancipated spectator*. London: Verso Books.

Rogoff, I. (2000). *Terra Infirma: Geography's visual culture* (Vol. 49). London: Psychology Press.

Save the Children. (2016). *Layla and Sana sleep without fear of snakes*. Lebanon: Save the Children. https://lebanon.savethechildren.net/news/layla-and-sana-sleep-without-fear-snakes

Save the Children. (2017). *Advocacy*. Save the Children International. https://www.savethechildren.net/advocacy

Sigona, N. (2013). The politics of refugee voices: Representations. In E. Fiddian-Qasmiyeh, G. Loescher, K. Long, & N. Sigona (Eds.), *The Oxford handbook of refugee and forced migration studies* (p. 369). New York: Oxford University Press.

Simms, N. (1986). *Silence and invisibility: A study of the literatures of the Pacific, Australia, and New Zealand*. Washington, DC: Three Continents Press.

Slingenberg, L., & Bonneau, L. (2017). (In) formal migrant settlements and right to respect for a home. *European Journal of Migration and Law, 19*(4), 335–369.

Solomon-Godeau, A., & Nochlin, L. (1991). *Photography at the dock: Essays on photographic history, institutions, and practices*. Minneapolis: University of Minnesota Press.

Sontag, S. (1973). *On photography* [Online]. Electronic Edition 2005. New York: RosettaBooks LLC.

Sontag, S. (2003). *Regarding the pain of others*. London: Hamish Hamilton.

Sparkes, A. G. (1994). Self, silence and invisibility as a beginning teacher: A life history of lesbian experience. *British Journal of Sociology of Education, 15*(1), 93–118.

The Independent. (2015). Refugees are going on hunger strike at the Hungarian border. *The Independent*, September 15. http://www.independent.co.uk/news/world/europe/refugee-crisis-asylum-seekers-start-hunger-strike-as-they-refuse-to-leave-hungarian-border-10502112.html

The Nation, Maria. (2016). Desperate Syrian refugees begin a hunger strike in Greece. *The Nation*, May 18. https://www.thenation.com/article/desperate-syrian-refugees-begin-hunger-strike-in-greece/

UN General Assembly. (1967). *Declaration on territorial asylum*. http://www.refworld.org/docid/3b00f05a2c.html. Accessed 22 May 2018.

Van Leeuwen, T. (2009). Critical discourse analysis. In *The international encyclopedia of language and social interaction*. Chichester: Wiley Blackwell.

Wæver, O. (2000). The EU as a security actor: Reflections from a pessimistic constructivist on Postsovereign security orders. In M. Kelstrup & M. Williams (Eds.), *International relations theory and the politics of European integration* (pp. 250–294). London: Routledge.

Wilde, O. (2015). *Salomé*. Peterborough: Broadview Press.

Yurchak, A. (2013). *Everything was forever, until it was no more: The last soviet generation*. Princeton: Princeton University Press.

Chapter 8
Conclusion

8.1 Transnational Humanitarian NGOs and Global Governance

The investigation of the four relief agencies' organizational models – undertaken by combining analysis of websites, strategic documents and policy guidelines with fieldwork and interviews with NGO staffers – has shown the different ways in which each organization works. Exploration of the different sectors of intervention has highlighted the different roles NGOs want to have not only in the lives of their beneficiaries but more generally in the governance system of their communities. As illustrated in Chap. 5, the spectrum of activities is quite wide. Save the Children focuses on education and child protection (mainly through psychosocial support) complementary advocacy to secure policy change to enable a better world for children; Oxfam prioritizes 'giving voice' to the voiceless, water and sanitation, psychosocial support, legal counselling, combined also with a vigorous advocacy and influencing program to create lasting solutions to injustice and poverty. CARE has a similar focus on voice and empowerment especially for women and girls. Its gender transformative approach informs its work on protection, responses to gender-based violence distribution of relief items, and, to a lesser extent, water and sanitation. As with Save the Children and Oxfam, CARE sets store by advocacy for policy reforms to end poverty and gender inequality. For its part, MSF operations focused on medical assistance, ranging from primary health care, surgery, mental health and psychosocial support, and medical evacuation. For MSF, belief in the power of *témoignage* has driven denunciations of those who hinder humanitarian action or divert aid and also critique of the wider disfunctionalities of the humanitarian system itself.

The analysis has highlighted that despite diverse backgrounds, foci, and sectors of intervention, Save the Children, Oxfam and CARE all embody the typical organization of what some scholars have defined as 'new humanitarianism.' They are all heavily dependent on government-based funding. This makes their relationship

© The Author(s) 2021
A. Massari, *Visual Securitization*, IMISCOE Research Series,
https://doi.org/10.1007/978-3-030-71143-6_8

with state politics and their ability to negotiate independence from the overarching priorities of states complex, to say the least. Moreover, the three NGOs (in common with many other INGOs and UN agencies) have embraced collaboration with private capital as a positive opportunity for the development of humanitarian action. This has given rise to doubts as to the level of independence of humanitarian action vis-a-vis the economic and commercial priorities that, by definition, inspire the private sector.

In all three cases, the humanitarian goal has been combined with long-term development objectives, a holistic approach ideally able to 'tackle the root causes' of deprivation and human rights abuses. These substantial transformative ambitions are paradigmatic of the three organizations and their desire to play an active role in global governance. This is indicated not only in their aim to influence policy and societal change at the national and international level, but also in the fact that these transformational objectives have become a defining character of their respective organizational models. A distinctive common feature is the consistent shift toward a rights-based approach that privileges the promotion and the protection of human rights instead of targeting 'only' people's basic needs. In this sense, the rights-based approach is more aligned with a form of comprehensive humanitarian action that is interested in the individual as a rights-holder whose needs can be tackled only via addressing the political, economic, religious and cultural factors that underpin prevention of enjoyment of rights. Save the Children, Oxfam and CARE indeed claim for themselves an important role in global governance. For the three organizations, shaping, changing and inspiring policy change in the societies within which they operate is intended as a fundamental part of their mission. Paradoxically, however, they all understand and present their humanitarian rights-based work as purely technical, through their focus on rights, considered as a neutral dimension, located outside any political sphere.

MSF presents and performs its role in almost total opposition to the other NGOs. It has chosen to focus on a purely humanitarian mandate (as opposed to combining this with development work typical of multi-mandate agencies), to largely limit interventions to medical assistance, and to provide aid on the basis of need, instead of rights. This organizational model is combined with a quasi-independence from institutional funds and a strict policy on which funds to accept from governments. In this sense, MSF has opted for a unique position. Not claiming a role in global governance, MSF frequently stresses its *only* has medical and emergency intentions. At the same time, however, the NGO is the only one of those studied that has repeatedly stated that politics is a crucial dimension of its humanitarian work

In general, the analysis of the four transnational humanitarian NGOs' operations and sectors of intervention in the Syrian displacement response has shown that, despite internal variations and differences, they all perform work with implicit important political consequences in the communities and societies in which they operate and beyond. Each specific way through which relief agencies provide emergency assistance not only shapes humanitarian governance but the way refugees are perceived and 'managed' in global governance dynamics. In this sense, however, humanitarian performance does not unfold in an intellectual vacuum. On the contrary, interviews with practitioners have confirmed academic studies that have highlighted elevated levels of organizational reflexivity. NGO staffers have invested great efforts in proactively assessing, and reflecting on, the role and impact of their work. My conversations with humanitarian and advocacy practitioners and analysis of their organizational intervention modalities have shown how relief agencies do not only reflect extensively on the relationship that they intend to have with politics and global governance, but also how these internal reflections inform changes over time in their strategies, policies and operational principles.

8.2   Transnational Humanitarian NGOs and Securitization

This book has also explored the role that transnational humanitarian NGOs play in global governance, particularly with regard to the securitization of the refugee issue, through the way they visually represent people on the move. Analysis of the relief agencies' photographic accounts of Syrian displacement in 2015–2016 has shown that they contribute to the securitization of the refugee issue in three different yet interrelated ways. These I have conceptualized around three analytical themes: *threatening*, *threatened* and *(in)visibilities*.

With the term *threatening*, I refer to how NGOs' images contribute to reproduce the mainstream securitized media and political accounts depicting refugees in terms of threats. In this category I have identified five different aesthetic patterns that support a representation of displaced Syrians as threatening, thereby reinforcing a discursive process of securitization. The visual analysis has highlighted how humanitarian organizations' pictures of long lines of people pointing visually inside a familiar territory, with no visual goal in front of them, evoking penetrating arrows, support the narrative of the invasion. Similarly, the now quite familiar photos of overcrowded boats produced by relief agencies have, despite the use of innovative and alternative visual elements, continued to depict 'boat people' as dehumanized and incumbent masses, as 'others' eliciting feelings of fear among the public. In general images that represent people on the move as an exceptional emergency support a rhetoric of crisis and invasion that has pervaded populist and media securitized discourses on global mobility in general, and the European refugee/migration 'crisis in particular. Representation of the threatening refugee also is reinforced by images that consistently represent Syrians in abstract terms, essentialized people on the move, with no individuality, agency, political claims, or personal trajectories. In a similar way, pictures that represent the 'other', in binary terms as someone alien from 'us', reproduce a narrative based on questions of luck rather than justice, fail to facilitate comprehension of the wider political context and also reinforce an understanding of refugees as an exception and a threat to the otherwise 'normal' order of things.

It is very important to note that, overall, images of this kind are a minority and do not constitute a dominant form of humanitarian representation. However, their presence, even if marginal, acquires significant meaning when inscribed into the wider context of communication around refugees. In this sense, humanitarian discourse is expected – both by the general public and by how relief agencies present themselves – to defy media and populist accounts consistently depicting people on the move in terms of threat, and rather present an alternative narrative, a more 'humanitarian' one. Against this background, I have shown how NGOs have, in this sense, failed to move away and challenge these securitized accounts. Indeed, how to a certain extent, they have contributed to their reproduction and reinforcement. In so doing, I am not arguing that humanitarian organizations are intentionally reproducing a depiction of refugees in terms of threat. On the contrary, I acknowledge their efforts to defy mainstream securitized discourses. However, the intrinsic polysemic character of any pictorial representation allows for multiple meanings to be present in a single image. The combination of visual semiotic resources contributing to the representation of Syrian refugees in terms of threat has highlighted that this specific meaning cannot be overlooked in exploring the way in which humanitarian NGOs interrelate with securitization processes around refugees.

At the same time, it is crucial to underline that in their evident attempt to challenge media and populist political discourses that have represented people on the move as a threat, NGOs have tended to privilege a depiction of refugees as, rather, the referent object of the threat, as *threatened*. This concept is strictly connected with the changes that have interested the humanitarian enterprise over the last three decades: a re-conceptualization of the notion of human security which has put the individual and his/her protection at the very centre of the international security system, and the progressive shift from a needs-based towards a rights-based approach that considers beneficiaries of assistance as rights-holders. This paradigm shift has gone hand in hand with substantial changes in NGOs' humanitarian visual communication. The typical image of present-day humanitarian crisis is no longer that of a starving baby, but that of a displaced child stripped by war of her/his right to education. The threatened analytical theme is, indeed, the most widespread aesthetic pattern of contemporary humanitarian representation.

This study has explored six different visual themes contributing to the representation of Syrian refugees as threat in order to demonstrate how this representation, while opposed to that of threatening individuals, is simultaneously just another form of securitization, whereby Syrians are depicted as infantilized and passive victims in need of external intervention.

Images of pity, quite typical of traditional humanitarian communication, are still present in today's visual material and show an over-representation of women that implies a sort of embodied vulnerability in the female subject. Moreover, and most importantly, the iconography of pity – strictly connected with the religious iconography of *pietas* – utilizes the language of urgency instead of justice, thereby calling for immediate interventions that can end the suffering while concealing the wider socio-economic and political causes of that distress. Likewise, vivid images of suffering and physical pain found in the NGOs' visual material produce similar effects and underline the asymmetrical distance between the viewer and the (suffering and passive) represented participant.

The analysis has also confirmed a depiction of Syrian refugees, extensively discussed in the literature in other refugee contexts, in terms of essentialized victims. The representation of Syrian refugees as helpless and passive victims is further reinforced by a clear over-representation of children, that by metonymically symbolizing the general humanitarian subject, infantilizes people on the move, reproducing a colonial and paternalistic account. A further contribution, though with a different aesthetic pattern, has similar effects. This is the visual narrative of the 'white hero' saving the 'distant other' that reinforces a dichotomous and asymmetrical understanding of human solidarity.

Finally, the analysis has focused on the aesthetic pattern of innocence largely used by NGOs in their effort to defy representations of refugees as threat. Beside the positive symbolic value attributed to innocence, these kinds of images also contribute to the reification of dichotomous categories of deserving and undeserving refugees that remain caught in a perverse dynamic in which the burden of proof of innocence falls back on them. At the same time, the exploration of NGOs' photographic production has also highlighted a new important trend of humanitarian visual representation, a quite widely used representation seeking to show beneficiaries with dignity and empowerment. This more positive illustration of people on the move as rights-holders, however, simultaneously contributes to their securitization. It does so by diluting their individuality through an essentialized depiction as humanitarian subjects, by shaping the protection authority that the humanitarian actors have over them and by constituting the humanitarian intervention as the most legitimized security strategy to address human insecurity.

These different forms of securitized representation – in terms of threatened humanitarian subject, regardless of whether illustrations are positive or negative, not only fail to fulfill the emancipatory promise of the rights-based human security framework, but also reinforce securitized discourses of humanitarian governance. In fact, by reinforcing a depiction of Syrian refugees with limited agency, that weakens people on the move political claims, they inscribe them into dependency and protection dynamics that require external intervention and management.

Finally, the analysis of what is visible and what is invisible in the transnational humanitarian NGOs' photographic representation of Syrian displacement – the analytical theme of *(in)visibilities* – has shown how visual absence can contribute to the securitization of the refugee issue. The exploration of the invisibilities has highlighted how topics that are important for Syrians on the move such as political support for particular factions (political symbols or gestures) and protests (individual and collective contestation) are completely absent from NGOs visual communication.

Their absence is explained by different reasons, including the intention to protect humanitarian beneficiaries from potential security repercussions and relief organizations' efforts to not actively reproduce accounts of threatening refugees. The invisibility of politics and protest has also the effect of silencing refugees and depriving them of political agency. The exploration of NGOs' photos has also highlighted how the various factors that are causing suffering in Syria and displacement are completely overlooked. The only images connected to the topic of threat are in fact those of destruction and devastation from which, however, the perpetrator, armed groups, bombs or weapons remain completely absent. This invisibility, equivalent to the linguistic omission of specific grammatical elements, is not just a stylistic choice. It has political implications insofar it produces an account that conceals the real causes of threat. The same goes for the visual absence of the various kind of threats that affect Syrians in Jordan and Lebanon since the photos did not include any of the protection issues mentioned in various emergency reports and by NGO staff. The only exception is the visual account of the top humanitarian concerns on the situation of Syrian refugees in Greece, where the picture contained visual clues as to the protection priorities underlined by relief organizations, such as detention and border closures.

The investigation of visual absence has also considered its auditory equivalent: silence. In this sense, the analysis has demonstrated how despite the significant effort made by NGOs to 'make refugees' voices heard', the mediatization of the process and the constant representation of people through the lenses of their humanitarian needs, have the effect of silencing refugees and reproducing an account that depicts them mostly, if not exclusively, in terms of humanitarian subjects with very limited agency and in need of external intervention.

8.2.1 On the Possibility of 'Good' Securitization

Drawing on the discussion above, it would be tempting to conclude that securitization can be seen as both a negative process (when refugees are depicted as threat), but also positive (when refugees are shown threatened for their own protection). Obviously, transnational humanitarian NGOs represent displaced Syrians people as threatened and in need of external intervention in order to mobilize financial and other support and solidarity. They operate within a competitive media environment in which they try to create an alternative narrative to the one proposed by populist and right-wing political discourses. However, even taking this into account, it is important to reflect whether the securitization of Syrian refugees is the most appropriate framing for the humanitarian cause or if it would other frameworks would be preferable. This study argues that although the two forms of securitizations – refugees as threat and refugees as threatened – are based on distinct, if not opposed, premises and although very different aesthetic registers are at play, both representations resonate with the logic of emergency and governmentality. Both depictions in fact reinforce a representation of people on the move as in need of management and control, given the extraordinariness of the situation. The findings therefore confirm what has been theoretically pointed out by a significant range of scholars who have explored the theoretical connections between humanitarianism and securitization (Aradau 2004; Huysmans and Squire 2009); with governance (Barnett 2013); biopolitics (Reid 2010; Rozakou 2012): the politics of life (Fassin 2007), and humanitarian interventions as part of an "emergency imaginary" (Calhoun 2004). It has also shown empirically for the first time how relief organizations play an active role in the securitization of the refugee issue.

This securitization framework is problematic for two main reasons. First of all because it implies the idea of emergency as a way to address crisis that does not take into consideration the dynamics producing an 'emergency'. As Calhoun has observed, this perspective "also reflects a distanced view on the global system, a view from nowhere or an impossible everywhere that encourages misrecognition of the actual social locations from which distant troubles appear as emergencies" (Calhoun 2004, 378). The problem is that the securitization of an issues also implies the failure to deal with it through politics (Möller 2007), thus disregarding the political (but also social and economic) dynamics that need to be addressed. This view is in obvious contrast with transnational NGOs' intention to mobilize support by raising awareness of distant suffering while aspiring to tackle 'root causes' of crises. In adopting the 'emergency imaginary', through both the threat and threatened representation, relief organizations contribute to inscribe Syrian refugees' in this securitized framework, one that fails to provide an alternative message and is largely at odds with declared aspirations. As different NGOs' strategic documents point out, to tackle the root causes of crises implies addressing questions of socio-economic discrimination and unequal power relations, aspects completely pushed into the background by an emergency scenario.

Secondly, the securitization of the refugee issue strongly resonates with discourses of migration management and control. Both when people are represented as threat and when they are represented as threatened, humanitarian actors are crucial to addressing and managing the situation. In this sense, also the threatened framework, in which refugees emerge as essentialized humanitarian subjects with limited agencies, contributes to a discourse of migration governance that reifies dichotomic and asymmetrical understandings of people and movements, depicting a group of people who need to be managed and another who charged with management. Clearly, this does not chime with relief organizations' aspirations in the context of refugee response: as the analysis has demonstrated, the governance of international mobility is hardly a humanitarian NGOs' priority. However, as Aradau (2004) has shown, the two seemingly antithetical humanitarian and securitization discursive regimes appear, rather, mutually constitutive when understood in terms of governmental processes.

Against this background, relief organizations' appeal to the rights-based framework, far from representing a drastic change from securitization discourses, also fails to offer an alternative way to deal with today's global mobility. The association of the two frameworks is not surprising even when one considers humanitarian NGOs' emphasis on human rights and the rights-based approach. As this study has highlighted, many NGOs' tend to emphasize the neutrality and rights-based technicality of their interventions. With the same exact logic, as Geiger and Pécoud have argued: "the very notion of 'management' is characterized by its apolitical and technocratic nature, and its popularity (to the detriment of other notions such as 'the politics of migration') is in itself a way of depoliticizing migration. Policies would not result from political choices, but from 'technical' considerations and informal decision-making processes on the most appropriate and successful way of addressing migration" (Geiger and Pécoud 2014, 11). Aradau has cogently observed that:

"if human rights have become the rights of those who are too weak or too oppressed to actualize and enact them, they are not 'their' rights. They are deprived of political agency; the only rights are our rights to practice pity and humanitarian interventions. Victims are therefore divorced from the very possibility of political agency, turned into spectral presences on the scene of politics. (…) The political agency of the marginalised and the excluded, the powerless and the silenced is thus either effaced or pathologised, expunged from the truly political claims and implications it should have" (Aradau 2004, 276).

The recent humanitarian shift toward a rights-based approach (with its visual representation of the rights-holder with dignity and empowerment) has failed to substantially alter the mechanism of a dichotomous and asymmetrical power relation between the empowered rights-holder, to whom very limited agency is recognized, and the external (humanitarian) intervention that is urgently required to address abuse and human (in)security. In this sense, this study has empirically demonstrated what has been argued at the conceptual (Huysmans and Squire 2009) and practice level (Bigo 2002) on the migration/security nexus by also showing the specific role that humanitarian NGOs play. The study has highlighted how visual humanitarian discourse contributes to reinforcing a securitized discourse around refugees that justifies exclusionary distinctions between desirable and undesirable people on the move (Huysmans and Squire 2009), and that this eventually contributes to the securitization of migration (Bigo 2002, 2; Huysmans et al. 2006).

The analysis has not indicated that NGOs consider the securitization framework as the most efficient way through which to mobilize support and raise funds. It seems to suggest that they are truly attempting to find an alternative, more 'humanitarian' discourse. Interviews with practitioners and analysis of their strategic documents have highlighted a genuine effort to portray individuality, integrity, agency, empowerment and successes. However, the study has also shown how transnational humanitarian NGOs' attempts to go beyond the securitization framework have not so far been sufficiently successful, indeed how this very framework entails a form of collateral damage, the loss of agency and disempowerment of Syrian displaced people. Securitized representation that aim to highlight the 'emergency' nature of the situation, even when framed in terms of protection, such as in the case of the threatened humanitarian subject, implies a loss of agency. Again, this is in contrast with efforts to empower and give voice to refugees.

A question arises from these reflections: how it is possible to represent Syrian displacement effectively (in order to mobilize support and raise awareness) without securitizing the object of the representation? It is difficult to deny that a depiction focusing on the threatened subject and his/her human security may be functional to NGOs' struggles for public attention and resources. However, the implications of a securitized account are not exactly within the purposes of many relief organizations. Even while their securitized framework is effective to fundraise and draw attention to 'emergencies' the same NGOs have expressed intent to give voice to the voiceless and empower beneficiaries. In this sense, the study has highlighted how securitized accounts hardly empower refugees. An alternative discourse on global mobility – one able to fulfill transnational humanitarian organizations aspirations – appears possible only by renouncing the human security frame and opting for an approach based on the recognition of the political agency of people on the move. The rights-based approach – which most humanitarian actors and some scholars (Goodwin-Gill 2001; Piper and Rother 2012) identify as potentially inspiring alternative frameworks – can only work if human rights are re-conceptualized around people's individual and collective political agency and not as a neutral and apolitical dimension. As Rancière has argued, attention to the individual is not sufficient to ensure her/his emancipation, unless accompanied by acknowledgment of the individual as a political member of the community (Rancière 1998). Providing a practical solution regarding the best ways to show individual and collective political agencies in the visual representation of refugees is beyond the scope of this study. The section which follows will highlight what are the best practices observed in NGOs' visual communication and how these have potential to offer a non-securitized alternative humanitarian message.

It could be asked that if refugees are represented in positive terms and with full agency, why would they need any support? The point here is not to advocate for an artificially cheerful representation of Syrian refugees – where people are always smiling and portrayed as being in charge – because this would be naive and disregard the suffering that invariably accompanies conflict and displacement. However, a depiction of Syrians on the move including a wider array of visual themes, emotions, everyday challenges, and political agency would open new possibilities. Such a representation could offer clues regarding the wider socio-economic context and political dynamics. More focus on everyday experiences of refugees would allow transnational humanitarian NGOs to offer a better understanding of the distant suffering in response to which they seek to mobilise, while enabling a comprehension of the situation beyond the emergency and governance framework that, by eventually disempowering refugees, is of little help either to them or NGOs' humanitarian cause.

8.3 Glimmers

After having illustrated the criticalities, it is equally important to turn to the possibilities for change. To use a visual metaphor, the multi-modal analysis has identified what I would like to call *glimmers*. It is possible to recognize minor alterations in humanitarian discourse and spaces of visibility that can potentiality produce an alternative humanitarian discourse. It would be presumptuous –and beyond the scope of this study –to here offer solutions to the various problematic aspects highlighted by the study (especially those linked to the complicated question of political agency). What is important is to underline what the analysis has identified as positive trends in humanitarian discourse around refugees and modes of visual expression.

Firstly, interviews with practitioners, knowledge gained in the field and exploration of relief organizations' strategic documents have revealed that transnational humanitarian NGOs are trying hard to go beyond a securitized framework. As noted, in recent decades the humanitarian enterprise has shown high levels of internal reflexivity. The emphasis that NGOs attach to considering and depicting beneficiaries with dignity, empowerment and agency is an important sign of the direction that relief organizations are attempting to move in. In this sense, each agency's communication guidelines and such joint initiatives as the *Code of Conduct on Images and Messages* (Concord 2006) demonstrate that relief organizations are perfectly aware of how images should, or should not, look like. They encourage visual representations that show 'solidarity', 'equality', and 'justice', and call for rejection of images that can to stereotype, sensationalize or discriminate. They endeavor to ensure people may speak for themselves aiming to improve public understanding of the complexities of the situation on the ground. The way that NGO managements internally sign-off and disseminate depictions are meant to control visual communication and acknowledge how wrong images can have wrong effects. So far, these noble intentions there have only produced minor changes in how refugees are represented – such as more individual portraits, smiling faces or confident gazes – but they have the potentiality to influence change.

By focusing specifically on visuality, this study has shown that the visual presence of a wider range of aesthetic themes could prove to be extremely useful in representing the experience of displacement with more attention to its complexity. The visual presence of the threat, weapons, bombs, law enforcement and the apparatus of border regimes, techniques of control and surveillance and fences are important visual clues that, when included in images, help illustrate interconnections between the humanitarian situation of the people represented and wider sociopolitical dynamics. When NGOs' pictures include these visual elements, as has been the case in some images discussed in the analytical chapters, the level of understanding of the situation of the image is particularly enhanced. When photos show the condition of distress together with the perpetrator, even its symbol, they are more likely to foster understanding of the context and reflection. As Sontag has affirmed, "the images that mobilize conscience are always linked to a given historical situation. The more general they are, the less likely they are to be effective" (Sontag 1973, 12). Similarly, the aesthetic patterns relating to everydayness, routine, and the various strategies through which people navigate the strange interplay between protracted 'emergency' and normalization – beyond the dominant rhetoric focused on humanitarian needs – can be very helpful in conveying the individuality, personal trajectories and diversity of experiences linked to displacement. Deploying a wider range of visual themes seems to be the best way to start representing refugees with a higher level of political agency. By visually including the causes of displacement, the perpetrators of violence, people's political engagement, their grievances, protest and contestation, the mundane, their networks of solidarity and coping mechanisms NGOs could depict displaced Syrians in a more complex way, deconstructing an essentialized and de-politicized representation of threatened victims.

Obviously, a single image cannot include too many visual elements. How is it then possible to overcome this technical and conceptual challenge? One way lies in the visual or conceptual connection of images with each other. This can be simply achieved by including them in a specific photo gallery, making explicit the leitmotiv linking them. The expressive potentiality of this kind of medium can be significative in broadening understanding of refugees' identities and experience beyond the humanitarian subject, thus offering wider information regarding the context of war and displacement. This can be achieved through photo galleries dedicated to various aspects of one individual's personal experience of displacement, a sort of photo documentary able to offer a representation of the refugee's identity beyond his/her immediate humanitarian need, which includes the causes of displacement. The analysis of the photo galleries of the four selected NGOs' has not shown examples of such specific detailed focus on individuals. However, it has highlighted some examples of reportage which offer multiple visual perspective, such as the case of Oxfam's *Migrants Winter Walk* (Oxfam 2016) or MSF's *Trapped at the EU Border* (MSF 2016b). There are also extremely interesting communication examples of interactive webpages that allow the user to navigate through the different connected sections, photos, and text. An example is MSF's *Stay Alive – The Route from Syria to Europe* website (MSF 2016a). Such kinds of communication project offer promise for they allow the inclusion of a wider set of visual topics to add complexity to the illustration and enhance understanding of what is happening on the ground. Although, of course, a certain level of abstraction and generalization is part of any form of visual representation which is intended as selective and interpretative action, the more themes and perspectives which can be added to the story the more comprehensive will be the depiction.

Capitalizing on these attempts to develop more complex and problematized visual narratives could allow NGOs' representation of refugees to move beyond an essentialist and simplistic depiction in terms of humanitarian subjects and thus ultimately provide a more comprehensive portrayal of the complex realities of refugee displacement. Through these more inclusive forms of communication, we might perhaps glimpse the glimmers of an alternative discourse, able to challenge dominant and securitized forms of discourse on people on the move.

References

Aradau, C. (2004b). The perverse politics of four-letter words: Risk and pity in the securitisation of human trafficking. *Millennium, 33*(2), 251–277.

Barnett. (2013). Humanitarian governance. *Annual Review of Political Science, 16*, 379–398.

Bigo, D. (2002). Security and immigration: Toward a critique of the governmentality of unease. *Alternatives, 27*(1_suppl), 63–92.

Calhoun, C. (2004). A world of emergencies: Fear, intervention, and the limits of cosmopolitan order. *Canadian Review of Sociology/Revue Canadienne de Sociologie, 41*(4), 373–395.

Concord. (2006). *Code of conduct on images and messages.* https://concordeurope.org/wp-content/uploads/2012/09/DEEEP-BOOK-2014-113.pdf?c676e3&c676e3

Fassin, D. (2007). Humanitarianism as a politics of life. *Public Culture, 19*(3), 499–520.

Geiger, M., & Pécoud, A. (2014). International organisations and the politics of migration. *Journal of Ethnic and Migration Studies, 40*(6), 865–887.

Goodwin-Gill, G. S. (2001). Refugees: Challenges to protection. *International migration review, 35*(1), 130–142.

Huysmans, J., & Squire, V. (2009). Migration and security. In *Handbook of security studies* (pp. 169–179). London: Routledge.

Huysmans, J., Dobson, A., & Prokhovnik, R. (Eds.). (2006). *The politics of protection: sites of insecurity and political agency* (Vol. 43). Psychology Press.

Möller, F. (2007). Photographic interventions in Post-9/11 security policy. *Security Dialogue, 38*(2), 179–196.

MSF. (2016a). *Stay alive*. 2016. http://stayingalive.msf.org/

MSF. (2016b). Trapped at Europe's borders by MSF. *Exposure*, 2016. https://msf.exposure.co/trapped-at-europes-borders

Oxfam. (2016). The migrants' winter walk by Oxfam International on Exposure. *Exposure*, 2016. https://oxfaminternational.exposure.co/the-migrants-winter-walk?locale=encategories%2Ffamilycategories%2Fbusinesscategories%2Ffamilycategories%2Ffamilycategories%2Fwedding scategories%2Fweddingscategoriescategories%2Fbusinesscategories%2Fbusiness%3Fmore%3Dtrue%3Fmore%3Dtrue?more=true

Piper, N., & Rother, S. (2012). Let's argue about migration: Advancing a right (s) discourse via communicative opportunities. *Third world wuarterly, 33*(9), 1735–1750.

Rancière, J. (1998). The cause of the other. *Parallax, 4*(2), 25–33.

Reid, J. (2010). The biopoliticization of humanitarianism: From saving bare life to securing the biohuman in post-interventionary societies. *Journal of Intervention and Statebuilding, 4*(4), 391–411.

Rozakou, K. (2012). The biopolitics of hospitality in Greece: Humanitarianism and the management of refugees. *American Ethnologist, 39*(3), 562–577.

Sontag, S. (1973). *On photography* [Online]. Electronic Edition 2005. New York: RosettaBooks LLC.

The manufacturer's authorised representative in the EU is Springer
Nature Customer Service Centre GmbH, Europaplatz 3, 69115 Heidelberg,
Germany. If you have any concerns regarding our products, please
contact ProductSafety@springernature.com

Printed and bound by CPI Group (UK) Ltd, Croydon, CR0 4YY
29/04/2026
02099458-0006